Get the most from this book

Welcome to your Revision Guide for the Pure Mathematics content of the MEI AS Level in Mathematics course (OCR Mathematics B). This book will provide you with reminders of the knowledge and skills you will be expected to demonstrate in the exam with opportunities to check and practice those skills on exam-style questions. Additional hints and notes throughout help you to avoid common errors and provide a better understanding of what's needed in the exam. In order to revise the Applied Mathematics (Mechanics and Statistics) content of the course, you will need to refer to My Revision Notes: OCR B (MEI) A Level Mathematics Year 1/AS (Applied).

The material in this bo sitting the MEI A Leve students may prefer to use the Revision Guide for the Pure Mathematics content of the MEI A Level Mathematics course (OCR Mathematics B) which covers all the pure mathematics needed for the exam.

Included with the purchase of this book is valuable online material that provides full worked solutions to all the 'Target your revision', 'Exam-style questions' and 'Review questions', as well as full explanations and feedback to each answer option in the 'Test yourself' multiple choice questions. The **online material** is available at www.hoddereducation.co.uk/myrevisionnotesdownloads.

Features to help you succeed

Target your revision

Use these questions at the start of each of the three sections to focus your revision on the topics you find tricky. **Short answers** are at the back of the book, but use the **worked solutions online** to check each step in your solution.

About this topic

At the start of each chapter, this provides a concise overview of its content.

Before you start, remember

A summary of the key things you need to know **before** you start the chapter.

Key facts

Check you understand all the key facts in each subsection. These provide a useful checklist if you get stuck on a question.

Worked examples

Full worked examples show you what the examiner expects to see in order to ensure full marks in the exam. The examples cover the full spectrum of the type of questions you can expect.

Hint

Expert tips are given throughout the book to help you do well in the exam.

Common mistakes

Your attention is drawn to typical mistakes students make, so you can avoid them.

Test yourself

Succinct sets of multiple-choice questions test your understanding of each topic. Check your **answers online**. The **online feedback** will explain any mistakes you made as well as common misconceptions, allowing you to try again.

Exam-style questions

For each topic, these provide typical questions you should expect to meet in the exam. **Short answers** are at the back of the book, and you can check your working using the **online worked solutions**.

Review questions

After you have completed each of the three sections in the book, answer these questions for more practice. **Short answers** are at the back of the book, but the **worked solutions online** allow you to check every line in your solution.

At the end of the book, you will find some useful information:

Exam preparation

Includes hints and tips on revising for the AS Mathematics exam, and details the exact structure of the exam papers.

Make sure you know these formulae for your exam

Provides a succinct list of all the formulae you need to remember and the formulae that will be given to you in the exam.

Please note that the formula sheet as provided by the exam board for the exam may be subject to change.

During your exam

Includes key words to watch out for, common mistakes to avoid and tips if you get stuck on a question.

My revision planner

SECTION 3

Go online for:

- **full worked solutions and answers to the Test yourself questions**
- **full worked solutions to all Exam-style questions**
- **full worked solutions to all Review questions**
- **full worked solutions to the Target your revision questions**

www.hoddereducation.co.uk/myrevisionnotesdownloads

SECTION 1

Target your revision (Chapters 1–4)

1 **Use the symbols $\Rightarrow$ $\Leftarrow$ and $\Leftrightarrow$**
P: $(x-3)(x-4)>0$ and Q: $x>4$.
Which of the following describes the complete relationship between P and Q?
P $\Rightarrow$ Q P $\Leftarrow$ Q P $\Leftrightarrow$ Q
(see page 2)

2 **Use proof by deduction**
Prove that the sum of the squares of any two consecutive even numbers is divisible by 4.
(see page 2)

3 **Use proof by exhaustion**
A **perfect number** is a positive integer that is equal to the sum of its factors (excluding itself).
6 is a perfect number since the factors of 6 are 1, 2 and 3 and 1 + 2 + 3 = 6.
Prove that 6 is the only perfect number which is less than 10.
(see page 2)

4 **Find a counter example to disprove a conjecture**
By finding a counter example, disprove the following statement.
$n^2-8n+15$ is positive for all integer values of n.
(see page 2)

5 **Use and manipulate surds**
Show that $\sqrt{48}+\sqrt{12}$ can be written in the form $a\sqrt{b}$ where a and b are as a small as possible.
(see page 6)

6 **Rationalise the denominator of a surd**
Simplify $\frac{2-\sqrt{3}}{2+\sqrt{3}}$.
(see page 6)

7 **Use the laws of indices**
Simplify $\frac{12a^2b^3c^4}{(2ab^2c)^2}$.
(see page 6)

8 **Understand negative and fractional indices**
Write $2^{-\frac{1}{2}}+\sqrt{2}+2^{\frac{3}{2}}$ in the form $k\sqrt{2}$.
(see page 6)

9 **Work with quadratic equations**
Use factorising to solve $6x^2-7x-3=0$.
Hence sketch the curve $y=6x^2-7x-3$.
(see page 10)

10 **Complete the square**
Write $y=x^2-4x-3$ in the form $y=(x+a)^2+b$.
(see page 14)

11 **Use the discriminant of a quadratic**
Find the value of k so that the equation $4x^2+kx+9=0$ has one repeated root.
(see page 14)

12 **Solve simultaneous equations where one is equation is quadratic**
Find the coordinates of the points where the line $y=2x+1$ intersects the curve $y^2=6x+7$.
(see page 19)

13 **Solve linear inequalities**
Solve $-6<3(1-2x)\leqslant 15$.
(see page 23)

14 **Solve quadratic inequalities**
Solve $x^2-2x-15>0$.
(see page 23)

15 **Represent inequalities graphically**
Show graphically the region represented by $x^2+3x-5\leqslant 2x+1$.
(see page 23)

Short answers on page 151
Full worked solutions online

CHECKED ANSWERS

Chapter 1 Problem solving

About this topic

This topic develops the problem solving skills you have already learned at GCSE and shows you how to construct a mathematical argument and logically reasoned proof.

Before you start, remember

- algebra and geometry from GCSE maths
- different types of number from GCSE maths.

Problem solving

REVISED

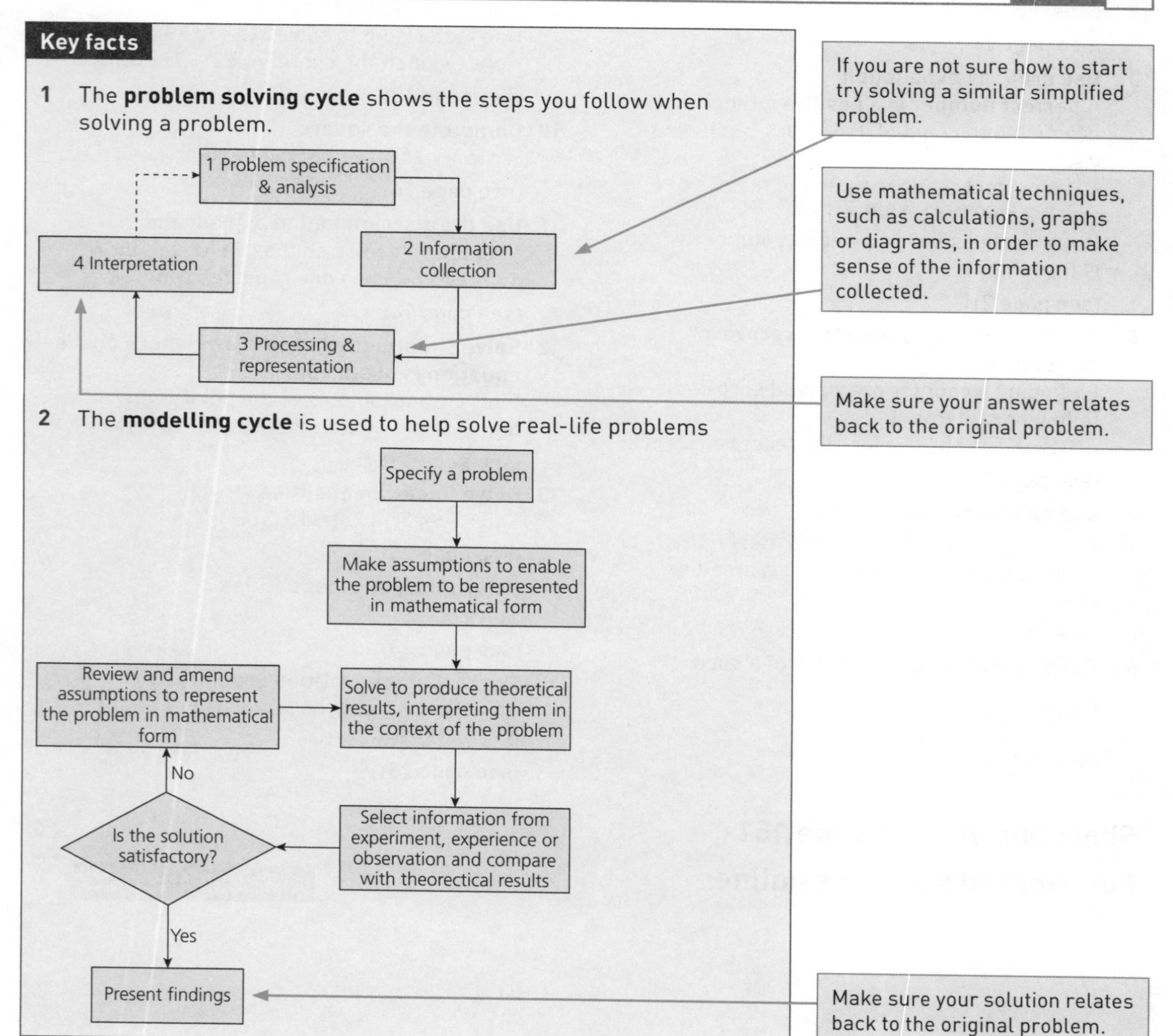

3 $P \Rightarrow Q$ means P **implies** Q or P **leads to** Q.

P is a **sufficient** condition for Q.

So Tom is a cat $\Rightarrow$ Tom is a mammal.

4 $P \Leftarrow Q$ means P is implied by Q or P follows from Q.

P is a necessary condition for Q.

Tom is a mammal does not imply Tom is a cat – he could be a dog!

5 $P \Leftrightarrow Q$ means P implies and is implied by Q or P is equivalent to Q.

P is a necessary and sufficient condition for Q.

For example, a number is even $\Leftrightarrow$ it is divisible by 2.

6 The **converse** of $P \Rightarrow Q$ is $P \Leftarrow Q$.

7 A **conjecture** is a mathematical statement which appears likely to be true, but has not been formally proved to be true.

You can prove a conjecture by:

- **Proof by exhaustion**
- **Proof by deduction.**

8 Sometimes it is easier to disprove a conjecture by finding a **counter example**.

So you can't say Tom is a cat $\Leftarrow$ Tom is a mammal.

You test every possible case – to exhaust all possibilities.

Start from a known result and then construct a logical argument as to why the conjecture must be true. This type of proof often uses algebra.

Worked examples

1 Using the symbols $\Rightarrow$, $\Leftarrow$ and $\Leftrightarrow$

Say whether or not the following statements are correct.

i The polygon is a quadrilateral $\Rightarrow$ the polygon is a square.

ii The polygon is a quadrilateral $\Leftarrow$ the polygon is a square.

iii The sum of the interior angles of the polygon is 360° $\Leftrightarrow$ the polygon is a quadrilateral.

Solution

i *This is not correct. A quadrilateral is any four-sided polygon, this does not imply that it is necessarily a square.*

ii *This is correct. All squares have four sides so 'the polygon is a quadrilateral' follows from the statement the polygon is a square.*

iii *This is correct. The sum of the interior angles of all four-sided shapes is 360° so the two statements are equivalent.*

For example, this quadrilateral is a parallelogram:

2 Finding the converse of a theorem

A theorem states that 'For triangle ABC, $CA = CB \Rightarrow \angle A = \angle B$'.

i State the converse of the theorem.

ii Is the theorem true? Is the converse true?

Solution

i For triangle ABC, $\angle A = \angle B \Rightarrow CA = CB$

ii Triangle ABC could be an isosceles triangle or an equilateral triangle.

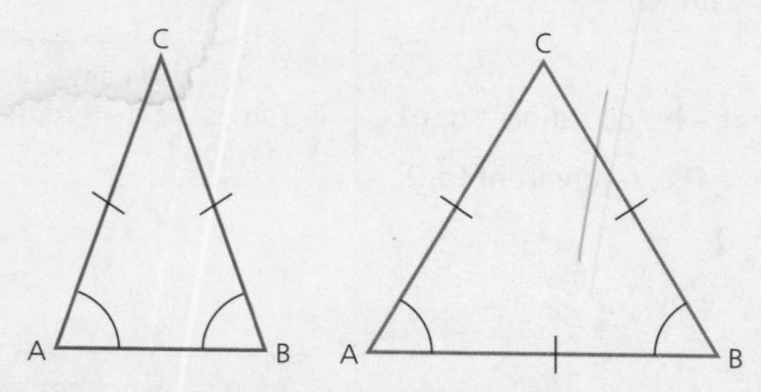

If the triangle is isosceles or equilateral then

$CA = CB \Rightarrow \angle A = \angle B$

and $\angle A = \angle B \Rightarrow CA = CB$

which can be written as $CA = CB \Leftarrow \angle A = \angle B$.

Hint: Since the theorem and its converse are both true, you can use the symbol '$\Leftrightarrow$': $CA = CB \Leftrightarrow \angle A = \angle B$.

Take care, it is quite often the case that a theorem is true but its converse is false.

3 Proof by deduction

Prove the conjecture 'the sum of two consecutive integers is always an odd number'.

Solution

Step 1: Let the first integer be n then the second integer is $n + 1$.

Step 2: Add the integers $n + (n + 1) = 2n + 1$.

Step 3: Complete the argument:

$2n$ is even since it is a multiple of 2
$\Rightarrow 2n + 1$ is odd since it is one more.

Hence the sum of two consecutive integers is always an odd number.

Hint: It does not matter whether n is odd or even, the next number will always be $n + 1$ and will be odd if n is even and even if n is odd.

4 Proof by exhaustion

Prove that square numbers with two digits are squares of numbers with one digit.

Solution

There are only six square numbers with two digits:

16, 25, 36, 49, 64, 81.

The square roots of these numbers are

$\pm4, \pm5, \pm6, \pm7, \pm8$, and ±9,

all of which are single digit numbers. Every case has been tested, so the conjecture must be true.

Hint: Proof by exhaustion is effective when there are only a limited number of cases to check.

Test yourself

TESTED

1

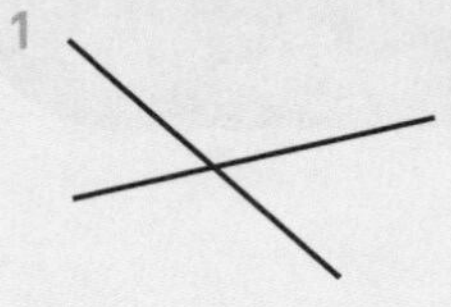
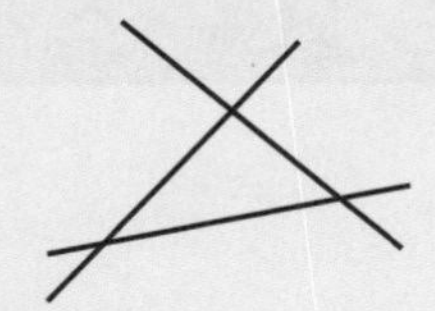
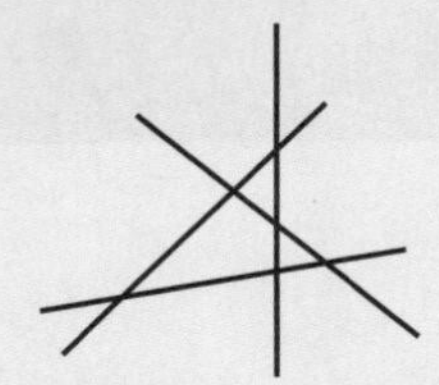

If you draw two lines, the greatest possible number of crossing points is one.
For three lines, the greatest possible number is three.
What is the greatest possible number of crossing points for n lines?

A $2n-3$ B $n(n-1)$ C $\frac{n^2-1}{2}$ D $\frac{n(n-1)}{2}$ E $3n-6$

2 Four of these statements are true and one is false. Which one is false?

A x is negative is a necessary condition for x^5 to be negative.

B For $ax^2+bx+c=0$ to have two distinct, real roots, $b^2-4ac \geqslant 0$.

C 3^n is an odd number for any positive integer n.

D The sum of any three consecutive odd numbers is divisible by 3.

E $0.\dot{3}$ is a rational number.

3 Look at the following statements about non-zero numbers, x and y. Which of them are true?

(1) $xy=1 \Rightarrow x=1, y=1$ (2) $xy=1 \Leftarrow x=\frac{1}{y}$ (3) $xy=1 \Rightarrow x=\frac{1}{y}$ or $y=\frac{1}{x}$

A (1) only B (2) only C (3) only D (1) and (3) E (2) and (3)

4 Consider the following statements:

P: n is an odd integer Q: n^2 is an odd integer R: n^2+n is an odd integer.

Only one of these gives a complete description of the relationship between the given pair of P, Q and R. Which one is it?

A $Q \Leftrightarrow P$ B $P \Rightarrow R$ C $Q \Rightarrow P$ D $R \Leftarrow Q$ E $P \Leftrightarrow R$

5 A proof consists of the lines marked A, B, C and D below. The proof may contain one or more errors. In which line is the first error. If there are no errors choose answer E.

Prove that for any real number, N, where $N \neq 0$, $N^0 = 1$.

A Let $N^0 = N^{(x-x)}$ B $\Rightarrow N^0 = N^x \div N^{-x}$ C $\Rightarrow N^0 = \frac{N^x}{N^x}$ D $\Rightarrow N^0 = 1$

Full worked solutions online

CHECKED ANSWERS

Exam-style question

The smallest of five consecutive integers is n.

i Write down the next four integers in terms of n.

ii Prove that the sum of any five consecutive integers is divisible by 5.

iii Using your result from part ii write down the sum of 17, 18, 19, 20, 21.

Short answers on page 151

Full worked solutions online

CHECKED ANSWERS

Chapter 2 Surds and indices

About this topic

An **irrational number** is a number that can't be written as a fraction $\frac{a}{b}$, where a and b are integers (whole numbers). For example, $\sqrt{2} = 1.414213\ldots$ is an irrational number as it can't be expressed as a fraction – the decimal part continues forever and never repeats.

Indices is the plural of **index**; index is another word for **power**. You will need to use the laws of indices to help you manipulate and simplify expressions.

Before you start, remember

- how to expand brackets
- how to simplify expressions
- how to use the laws of indices from GCSE maths.

Surds and indices

REVISED

Key facts

1. A **surd** is an expression containing an irrational root, such as $5 + \sqrt{3}$ or $2 - \sqrt[3]{7}$.
2. A surd is in its **simplest form** when the number under the square root has no square factors.
   - $\sqrt{20}$ is not in simplest form
   - $2\sqrt{5}$ is in simplest form.
3. You can **add and subtract surds** to **simplify them** in the same way as algebraic expressions.
4. When you multiply surds, remember:
   $\sqrt{x} \times \sqrt{x} = x$ $\quad \sqrt{xy} = \sqrt{x}\sqrt{y}$.
5. When a fraction has a surd in the denominator it is **not** in its simplest form.
   You simplify it by **rationalising the denominator**.
   For fractions in the form:
   - $\frac{1}{\sqrt{a}}$ multiply the top and bottom lines by $\sqrt{a}$
   - $\frac{1}{a + \sqrt{b}}$ multiply the top and bottom lines by $a - \sqrt{b}$.

   Remember $(a + b)(a - b) = a^2 - b^2$
   and $\left(\sqrt{a} + \sqrt{b}\right)\left(\sqrt{a} - \sqrt{b}\right) = a - b$.
6. In the expression a^m, a is the **base** and m is the **index** or **power** to which the base is raised.

Keep the rational numbers and the square roots separate.

Remember $\sqrt{}$ means the positive square root only.

7 The **laws of indices** are:
- $a^m \times a^n = a^{m+n}$ — Multiplication.
- $\frac{a^m}{a^n} = a^{m-n}$ — Division.
- $(a^m)^n = a^{mn}$. — Power of a power.

8 Remember that any non-zero number to the **power zero** is equal to 1.
$5^0 = 1$ $(-3)^0 = 1$ $1.7^0 = 1$ $a^0 = 1$ — 0^0 is undefined.

9 For negative and fractional powers
- $a^{-m} = \frac{1}{a^m}$ — A negative index indicates a reciprocal.
- $a^{\frac{1}{m}} = \sqrt[m]{a}$ — A fractional index is a root.
- $a^{\frac{m}{n}} = \sqrt[n]{a^m}$.

Worked examples

1 Simplifying expressions involving indices

Simplify:

i $3a^5 \times 4a^2$ ii $\frac{39x^4}{13x^3}$

iii $(5y^4)^2$ iv $2a^3b^2 \times 4a^2b^4$.

Solution

i $3a^5 \times 4a^2 = 3 \times 4 \times a^5 \times a^2 = 12a^{5+2} = 12a^7$

ii $\frac{39x^4}{13x^3} = \frac{39}{13} \times \frac{x^4}{x^3} = 3x^{4-3} = 3x$

iii $(5y^4)^2 = 5^2 \times (y^4)^2 = 25y^{4\times 2} = 25y^8$

iv $2a^3b^2 \times 4a^2b^4 = 2 \times 4 \times a^3 \times a^2 \times b^2 \times b^4 = 8a^5b^6$

2 Using index notation (1)

Simplify $27^{\frac{2}{3}}$.

Solution

$27^{\frac{2}{3}} = \left(27^{\frac{1}{3}}\right)^2 = 3^2 = 9$

Hint: It is also possible to do this calculation like this
$27^{\frac{2}{3}} = (27^2)^{\frac{1}{3}} = 729^{\frac{1}{3}} = 9$
But this would be harder to work out.

3 Using index notation (2)

Write $\frac{1}{81}$ in the form 3^n.

Solution

$81 = 3^4$

So $\frac{1}{81} = \frac{1}{3^4} = 3^{-4}$

4 Solving equations

Solve the equation $\frac{(2^x)^2}{2} = \sqrt[3]{2}$.

Solution

$$\frac{(2^x)^2}{2} = \frac{2^{2x}}{2} = 2^{2x-1} \text{ and } \sqrt[3]{2} = 2^{\frac{1}{3}} \Rightarrow 2^{2x-1} = 2^{\frac{1}{3}}$$

Start by writing both sides as a single power of 2.

Equating indices gives:

$$2x - 1 = \frac{1}{3}$$

So $2x = \frac{4}{3} \Rightarrow x = \frac{2}{3}$

5 Simplifying more complicated expressions

Simplify $\frac{\left(36^2 \times \frac{1}{2^3} \times \sqrt{3}\right)^4}{6^3}$.

Solution

$$\frac{\left(36^2 \times \frac{1}{2^3} \times \sqrt{3}\right)^4}{6^3} = \frac{\left((2^2 \times 3^2)^2 \times 2^{-3} \times 3^{\frac{1}{2}}\right)^4}{(2 \times 3)^3}$$

$$= \frac{\left(2^4 \times 3^4 \times 2^{-3} \times 3^{\frac{1}{2}}\right)^4}{2^3 \times 3^3}$$

$$= \frac{\left(2^{4-3} \times 3^{4+\frac{1}{2}}\right)^4}{2^3 \times 3^3} = \frac{\left(2 \times 3^{\frac{9}{2}}\right)^4}{2^3 \times 3^3}$$

$$= \frac{2^4 \times 3^{18}}{2^3 \times 3^3} = 2 \times 3^{15}$$

Hint: When a question involves using more than one base, you can split them up using the rule

$(a \times b)^n = a^n \times b^n$

This can help you simplify the solution to more complicated problems.

6 Simplifying surds

Simplify:

i $\sqrt{32}$ ii $4\sqrt{3} + 3\sqrt{3}$ iii $(4 + 2\sqrt{11}) + (5 - 6\sqrt{11})$.

Solution

i $\sqrt{32} = \sqrt{16 \times 2} = \sqrt{16} \times \sqrt{2} = 4\sqrt{2}$

ii $4\sqrt{3} + 3\sqrt{3} = 7\sqrt{3}$

iii $(4 + 2\sqrt{11}) + (5 - 6\sqrt{11}) = 9 - 4\sqrt{11}$

Hint: To simplify a surd look for factors that are square numbers.

7 Expanding brackets

Expand $(3 + 3\sqrt{7})(2 - 4\sqrt{7})$.

Solution

$$(3 + 3\sqrt{7})(2 - 4\sqrt{7}) = 3(2 - 4\sqrt{7}) + 3\sqrt{7}(2 - 4\sqrt{7})$$

$$= 3 \times 2 - 3 \times 4\sqrt{7} + 3\sqrt{7} \times 2 - 3\sqrt{7} \times 4\sqrt{7}$$

$$= 6 - 12\sqrt{7} + 6\sqrt{7} - 12 \times 7$$

$$= -78 - 6\sqrt{7}$$

8 Rationalising the denominator (1)

Simplify $\frac{5}{\sqrt{3}}$.

This is not in the 'simplest form' because the bottom line is a surd.

Solution

$$\frac{5}{\sqrt{3}} = \frac{5}{\sqrt{3}} \times \frac{\sqrt{3}}{\sqrt{3}} = \frac{5\sqrt{3}}{3}$$

9 Rationalising the denominator (2)

Rationalise the denominator $\frac{5-2\sqrt{3}}{4+2\sqrt{3}}$.

Solution

$$\frac{5-2\sqrt{3}}{4+2\sqrt{3}} = \frac{5-2\sqrt{3}}{4+2\sqrt{3}} \times \frac{4-2\sqrt{3}}{4-2\sqrt{3}}$$

$$= \frac{20-10\sqrt{3}-8\sqrt{3}+12}{4^2-(2\sqrt{3})^2}$$

$$= \frac{32-18\sqrt{3}}{16-12} = \frac{32-18\sqrt{3}}{4} = \frac{16-9\sqrt{3}}{2}$$

Multiply top and bottom by $4-2\sqrt{3}$ to make the bottom line a whole number.

Common mistake: You must multiply both the top and bottom by the same thing (so you are really multiplying by 1), otherwise you will change the value of the fraction.

Test yourself

TESTED

Make sure you can work these out without using a calculator!

1 Find the value of $\left(\frac{1}{3}\right)^{-2}$.

A -9 B $\frac{1}{9}$ C 9 D $-\frac{1}{9}$ E $-\frac{2}{3}$

2 Find the value of $\frac{36^{\frac{1}{2}}}{16^{\frac{3}{4}}}$, giving the answer in its simplest form.

A $\frac{3}{4}$ B $\frac{3}{2}$ C $\left(\frac{36}{16}\right)^{-\frac{1}{4}}$ D 3 E $\frac{6}{8}$

3 Simplify $(2-2\sqrt{3})$, giving your answer in factorised form.

A 16 B $8(2-\sqrt{3})$ C $-8(1+\sqrt{3})$ D $16-8\sqrt{3}$ E $4(4-\sqrt{3})$

4 Simplify $\frac{(2x^4y^2)^3}{10\left(x^3\sqrt{y^5}\right)^2}$.

A $\frac{1}{5}x^6y$ B $\frac{2}{25}x^6y$ C $\frac{4x^6}{5y^4}$ D $\frac{4}{5}x^6y$ E $\frac{8x^{12}y^6}{10x^6y^5}$

5 Find the exact answer to $\sqrt{54\times 48}$ simplifying your answer as much as possible.

A 50.9 B $36\sqrt{2}$ C $12\sqrt{18}$ D 36 E $\sqrt{2592}$

Full worked solutions online

CHECKED ANSWERS

Exam-style question

In this question you must show detailed reasoning.

Find the value of a and of b in each of the following cases.

i $3^{a-2}\times 5^{2b-1} = \frac{1}{\sqrt{5}}$

ii $\frac{5\sqrt{5}-\sqrt{2}}{\sqrt{5}+\sqrt{2}} = a-2\sqrt{b}$.

Short answers on page 151

Full worked solutions online

CHECKED ANSWERS

Chapter 3 Quadratic functions

About this topic

Quadratic equations appear in many situations throughout mathematics. It is important to be able to solve them confidently by factorising and also to be able to sketch the graph of a quadratic function.

Completing the square and the quadratic formula give further techniques for solving quadratic equations. You can also use completing the square to find the vertex and line of symmetry of the graph of a quadratic function. The quadratic equation formula is based on completing the square and is used to solve quadratic equations which can't be factorised.

Before you start, remember

- how to expand brackets
- how to factorise expressions.

Quadratic equations

REVISED

Key facts

1 A **quadratic equation** can be written in the form $ax^2 + bx + c = 0$ where $a \neq 0$ and b and c can be any number.

b and c can be positive, negative or zero.

2 Some quadratic equations can be solved by **factorising**.

See **examples 1** and **2** for a reminder of this.

Remember:

- a **perfect square** is an expression in the form $x^2 + 2ax + a^2 = (x + a)^2$
- the **difference of two squares** is an expression in the form $x^2 - a^2 = (x + a)(x - a)$.

3 The graph of a quadratic equation is a curve called a **parabola**.

To sketch the quadratic equation $y = ax^2 + bx + c$

- look at the sign of the x^2 term as this tells you which way up the curve is

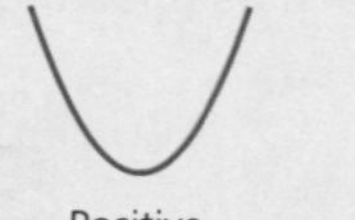

- find where the curve cuts the x-axis by solving $ax^2 + bx + c = 0$
- remember the curve cuts the y-axis at c.

When $x = 0$ then $y = c$.

Worked examples

1 Factorising a quadratic expressions where the coefficient of x^2 is 1

Factorise $x^2 + 4x - 21$.

Remember x^2 means $1x^2$.

Solution

Step 1: Find the product of the two outside numbers: -21.

$1 \times (-21) = -21$.

Step 2: Look for two numbers which multiply to give -21 and add to give 4.

The middle term is $+4x$.

These numbers are 7 and -3.

Step 3: Split the middle term using these numbers:

$$x^2 + 4x - 21 = x^2 + 7x - 3x - 21$$

Step 4: Factorise in pairs:

$$= x(x+7) - 3(x+7)$$
$$= (x-3)(x+7)$$

The two terms have a common factor of $(x + 7)$.

2 Factorising a quadratic expression where the coefficient of x^2 is not 1

Factorise $3x^2 - 20x + 12$.

Solution

Step 1: Find the product of the two outside numbers: 36.

$3 \times 12 = 36$.

Step 2: Look for two numbers which multiply to give 36 and add to give -20.

The middle term is $-20x$.

These numbers are -2 and -18.

Step 3: Split the middle term:

$$3x^2 - 20x + 12$$
$$= 3x^2 - 2x - 18x + 12$$

Step 4: Factorise in pairs:

$$= x(3x-2) - 6(3x-2)$$
$$= (x-6)(3x-2)$$

There is now a common factor of $(3x - 2)$.

3 Solving a quadratic equation

Solve the equation $2x^2 - 5x - 12 = 0$.

Solution

Factorising the left-hand side:

$$2x^2 - 5x - 12 = 0$$
$$2x^2 - 8x + 3x - 12 = 0$$
$$2x(x-4) + 3(x-4) = 0$$
$$(2x+3)(x-4) = 0$$

Split the middle term to help you factorise. First multiply the coefficient of x^2 and the constant: $2 \times (-12) = -24$.

Two numbers which add to give -5 and multiply to give -24 are -8 and $+3$. So you can split the middle term, $-5x$, into $-8x + 3x$.

Either $2x + 3 = 0$ or $x - 4 = 0$

So the solution is $x = -\frac{3}{2}$ or $x = 4$.

Since the product of the two factors is zero, one or other of them must equal zero, giving the two roots for the equation.

4 Sketching the graph of a quadratic

Sketch the graph of:

i $y = x^2 - 8x + 16$ ii $y = 25 - 4x^2$.

Solution

i When $x = 0$ then $y = 16$

When $y = 0$ then $x^2 - 8x + 16 = 0$

$\Rightarrow \quad (x - 4)^2 = 0$

$\Rightarrow \quad x = 4$

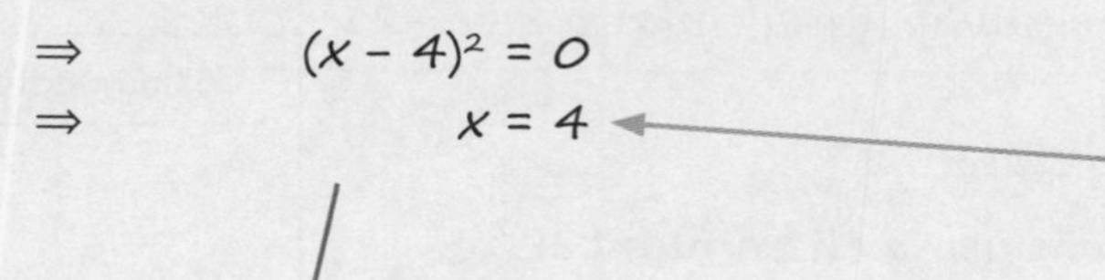

This is a **perfect square** as it is in the form $x^2 - 2ax + a^2 = (x - a)^2$.

There is one repeated root, so the curve just touches the x-axis.

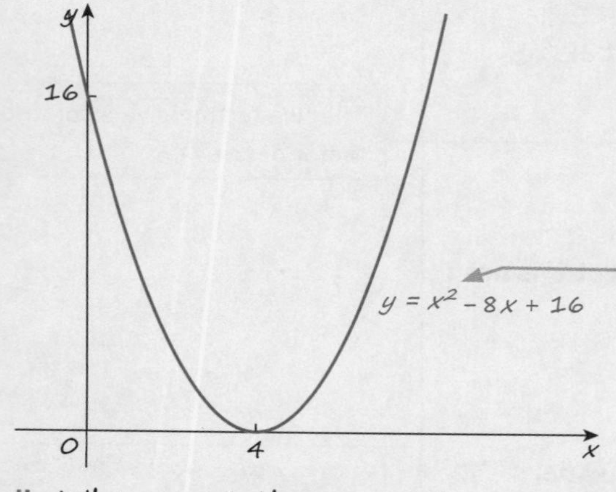

The coefficient of x^2 is positive so the curve is ∪-shaped.

ii When $x = 0$ then $y = 25$

When $y = 0$ then $25 - 4x^2 = 0$

$\Rightarrow (5 - 2x)(5 + 2x) = 0$

$\Rightarrow \quad x = 2.5$ or $x = -2.5$

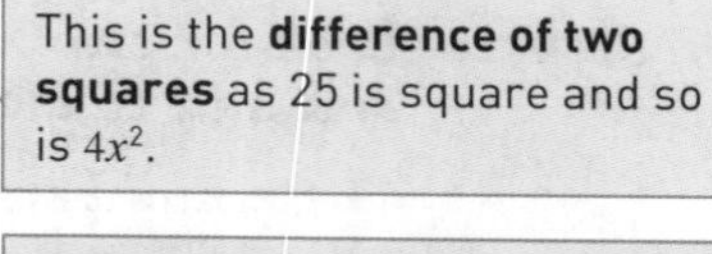

This is the **difference of two squares** as 25 is square and so is $4x^2$.

The coefficient of x^2 is negative so the curve is ∩-shaped.

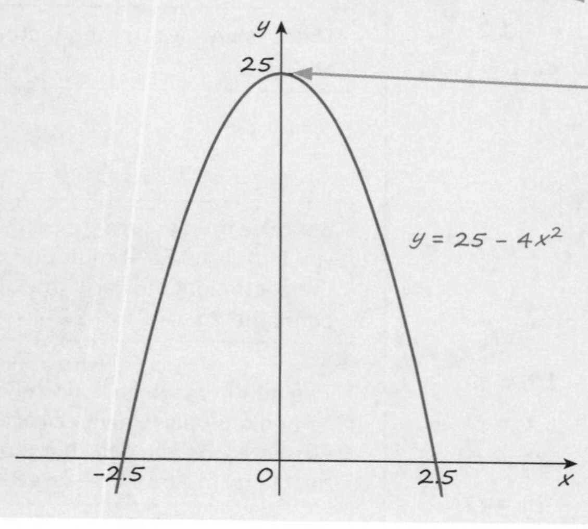

Don't forget to label where the curve crosses the y-axis.

5 Using quadratic equations

Solve $x - 7\sqrt{x} + 12 = 0$.

Hint: Some equations can be rewritten as a quadratic equation.

Solution

Replacing $\sqrt{x}$ with z gives the quadratic equation

$z^2 - 7z + 12 = 0$

$\Rightarrow \quad (z - 3)(z - 4) = 0$

$\Rightarrow \quad z = 3$ or $z = 4$

$\Rightarrow \quad \sqrt{x} = 3$ or $\sqrt{x} = 4$

$\Rightarrow \quad x = 9$ or $x = 16$

Remember $\left(\sqrt{x}\right)^2 = x$.

Common mistake: You need to solve the equation for x not z, so don't forget to work out what x is.

Test yourself

TESTED

Make sure you can work these out without using a calculator!

1 Which of the following is the solution of the quadratic equation $x^2 - 5x - 6 = 0$?

A $x = -6$ or $x = 1$ B $x = -2$ or $x = 3$ C $x = 2$ or $x = 3$ D $x = -3$ or $x = 2$ E $x = -1$ or $x = 6$

2 Factorise $6x^2 + 19x - 20$

A $(x + 4)(6x - 5)$ B $(3x + 10)(2x - 2)$ C $(x + 20)(6x - 1)$ D $(3x + 4)(2x - 5)$ E $(6x + 10)(x - 2)$

3 Which of the following is the solution of the quadratic equation $2x^2 - 9x - 18 = 0$?

A $x = \frac{3}{2}$ or $x = -6$ B $x = \frac{9}{2}$ or $x = -2$ C $x = 6$ or $x = 3$ D $x = -\frac{9}{2}$ or $x = 2$ E $x = -\frac{3}{2}$ or $x = 6$

4 Simplify the expression $\frac{x^2 - 9}{x^2 - x - 12}$ as far as possible.

Hint: Factorise the top and bottom lines. Then cancel any common factors.

A $\frac{9}{x - 12}$ B $\frac{3}{4}$ C $\frac{x + 3}{x - 4}$

D $\frac{x - 3}{x - 4}$ E $\frac{x + 3}{x + 4}$

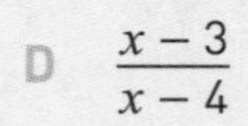

5 Find the equation of this curve.

A $y = 2x^2 + 3x - 9$

B $y = 2x^2 - 3x - 9$

C $y = -2x^2 + 9x - 9$

D $y = -2x^2 - 3x + 9$

E $y = 2x^2 + 9x + 9$

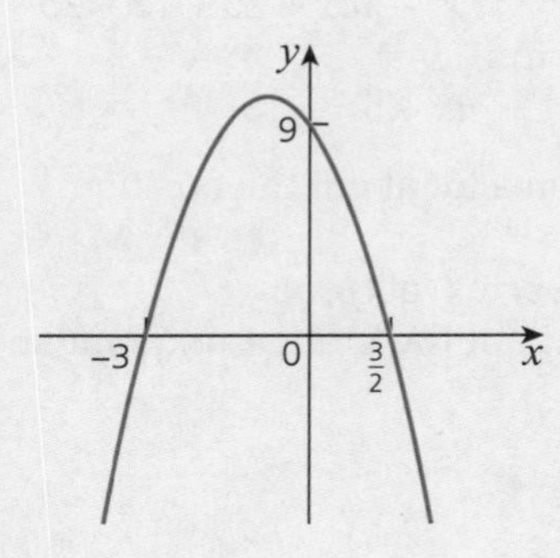

Full worked solutions online

CHECKED ANSWERS

Exam-style question

i Sketch the graph of $y = 4x^2 - 17x + 4$.

ii Solve:

a $4x^4 - 17x^2 + 4 = 0$

b $4x - 17\sqrt{x} + 4 = 0$.

Short answers on pages 151–2

Full worked solutions online

CHECKED ANSWERS

Completing the square and the quadratic formula

REVISED

Key facts

1 **Completing the square** means writing a quadratic expression in the form $a(x-p)^2+q$.
Completing the square is used to
- solve quadratic equations
- find the coordinates of the **vertex** (the turning point or stationary point) of the curve $y=a(x-p)^2+q$
- find the equation of the **line of symmetry** of the curve $y=a(x-p)^2+q$.

2 **Completing the square method.**

	Example $x^2+10x+12$.
• Halve the coefficient of x.	Half of 10 is 5
• Square it.	5^2 is 25
• Add it to the first two terms and subtract it from the constant term.	$x^2+10x+25+12-25$
• Factorise the first three terms, to make a perfect square.	$(x+5)^2-13$

You can only use this method when the coefficient of x^2 is 1. See **example 2**.

3 Completing the square tells you a lot about the position of the graph $y=(x-p)^2+q$.
- It has a **stationary point** (or vertex) at (p, q)
- The curve is symmetrical about the stationary point so the line $x=p$ is a **line of symmetry**.

4 The **quadratic formula** is

$$x=\frac{-b\pm\sqrt{b^2-4ac}}{2a}.$$

You can use the quadratic formula to solve quadratic equations written in the form $ax^2+bx+c=0$.

Hint: You should use the quadratic formula to solve equations which can't be factorised.

5 The **discriminant** is b^2-4ac.
The sign of the discriminant tells you how many real roots to expect.

Discriminant	Positive $b^2-4ac>0$	Zero $b^2-4ac=0$	Negative $b^2-4ac<0$
Number of real roots	2	1 (repeated)	0

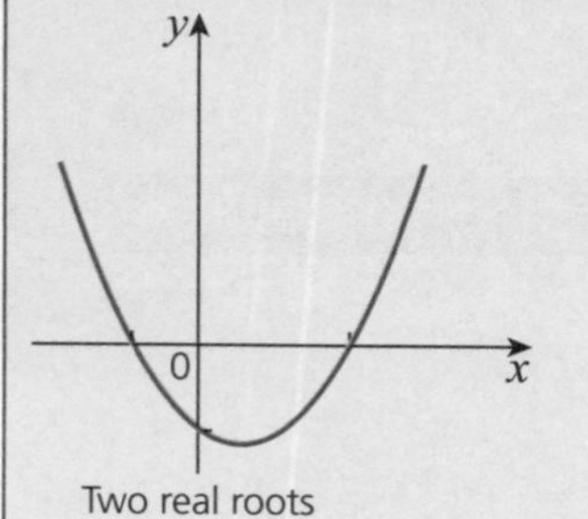

Two real roots

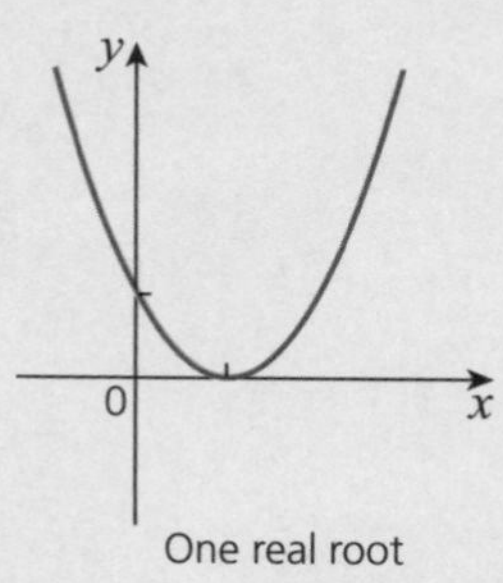

One real root

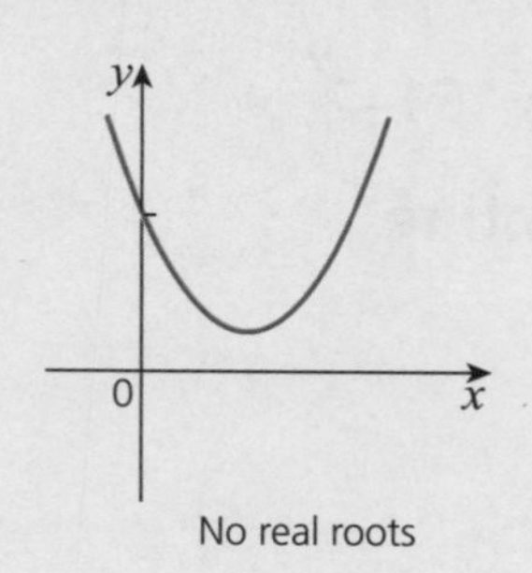

No real roots

Worked examples

1 Writing a quadratic expression in the form $(x+p)^2+r$

Write x^2+6x+4 in completed square form.

Solution

$$x^2+6x+4$$
$$= x^2+6x+9+4-9$$
$$= (x+3)^2+4-9$$
$$= (x+3)^2-5$$

Half of 6 is 3 and $3^2 = 9$ ...

... add 9 and then take it away so nothing has changed.

2 Writing a quadratic expression in the form $a(x+p)^2+r$

Write $2x^2-12x+5$ in completed square form.

Hint: When **completing the square,** the coefficient of x^2 must be 1. If it isn't you must start by taking out the coefficient of x^2 as a factor.

Solution

$$2x^2-12x+5$$
$$= 2[x^2-6x]+5$$
$$= 2[x^2-6x+9]+5-2\times 9$$
$$= 2(x-3)^2-13$$

Take 2 out as a factor of the two left hand terms.

Half of -6 is -3 and $(-3)^2 = 9$. Add 9 inside the bracket to make a perfect square ... then take it away so nothing has changed.

3 Using completing the square to sketch a curve (1)

Sketch the curve $y = x^2+6x+4$.

Solution

In **example 1** it was shown that the equation can be written as $y = (x+3)^2 - 5$, so the minimum point is at $(-3, -5)$ and the line of symmetry is $x = -3$.
From the equation, it can be seen that the curve cuts the y-axis at $(0, 4)$.
Find where the curve crosses the x-axis:

When $y = 0$, $(x+3)^2 = 5$

$$x+3 = \pm\sqrt{5}$$
$$x = -3 \pm\sqrt{5}$$

Substitute $x = 0$ into $y = x^2+6x+4$.

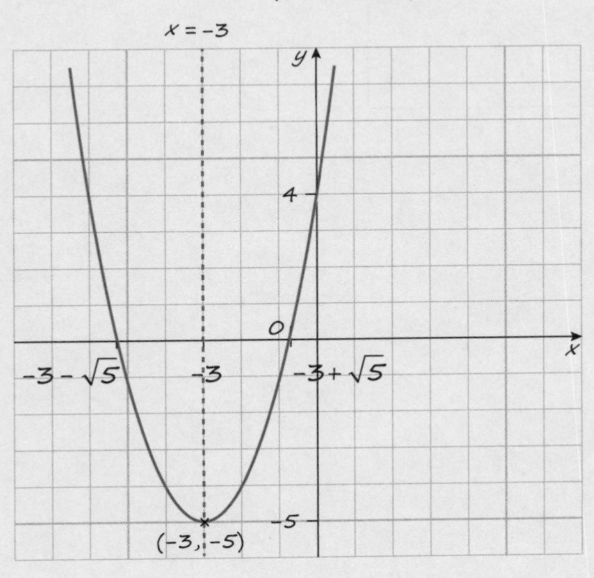

4 Using completing the square to sketch a curve (2)

Use completing the square to sketch the curve $y = -x^2 - 4x + 3$.

Solution

Completing the square gives

$$y = -x^2 - 4x + 3$$
$$\Rightarrow y = -(x^2 + 4x) + 3$$
$$\Rightarrow y = -(x^2 + 4x + 4) + 3 + 4$$
$$\Rightarrow y = -(x + 2)^2 + 3 + 4$$
$$\Rightarrow y = -(x + 2)^2 + 7$$

Take care! The negative sign outside the bracket means you have '-4' so you need '+4' to cancel this out.

The curve cuts the y-axis at $(0, 3)$.

The maximum point is $(-2, 7)$.

The line of symmetry is $x = -2$.

Find where curve crosses x-axis:

When $y = 0$, $(x + 2)^2 = 7$

$x + 2 = \pm\sqrt{7}$

$x = -2 \pm\sqrt{7}$

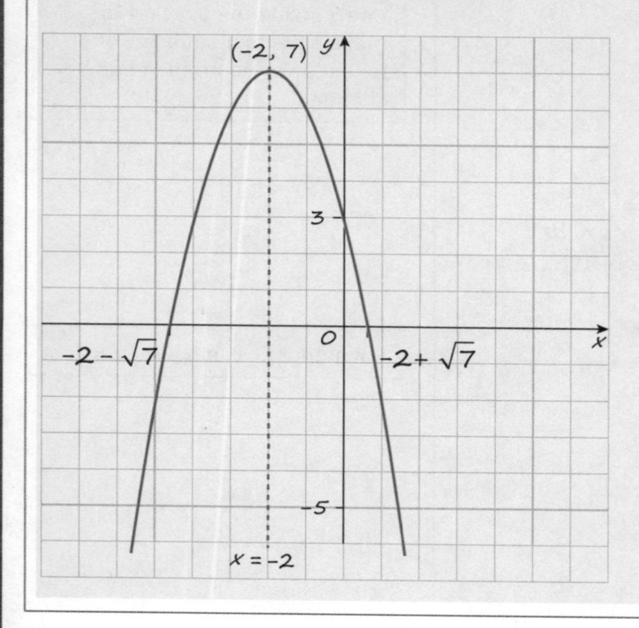

Common mistake: Don't forget that a curve with a negative coefficient for x^2 is upside down.

5 Using the quadratic formula

Solve the equation $5x^2 - 2x - 1 = 0$, giving your answers in **exact** form.

Hint: Exact means leave your answer as a surd or a fraction – not a rounded decimal.

Solution

In this case $a = 5$, $b = -2$ and $c = -1$.

Using the quadratic formula:

$$x = \frac{-(-2) \pm \sqrt{(-2)^2 - 4 \times 5 \times -1}}{2 \times 5}$$

$$= \frac{2 \pm \sqrt{24}}{10}$$

$$= \frac{2 \pm 2\sqrt{6}}{10}$$

$$= \frac{1 \pm \sqrt{6}}{5}$$

Common mistake: Take care with your signs!

Simplify the surd: $\sqrt{24} = \sqrt{4 \times 6} = \sqrt{4}\sqrt{6} = 2\sqrt{6}$

Common mistake: The rounded solutions are $x = 0.690$ or $x = -0.290$ (to 3 s.f.), but you aren't asked for these!

6 Using the discriminant

The equation $25x^2 + kx + 4 = 0$ has two real roots.

Find the possible values of k.

Hint: Use the discriminant for questions about number of roots.

Solution

When there are two real roots the discriminant $b^2 - 4ac > 0$

From the equation $25x^2 + kx + 4 = 0$

$a = 25, b = k$ and $c = 4 \Rightarrow k^2 - 4 \times 25 \times 4 > 0$

$\Rightarrow k^2 - 400 > 0$

$\Rightarrow k^2 > 400$

$\Rightarrow k < -20$ or $k > 20$

Common mistake: Take care with the inequality signs. Check values of k if you aren't sure which way round the signs should go.

Test yourself

TESTED

1 Which of the following is the solution of the quadratic equation $2x^2 - 3x - 4 = 0$?

A $x = \frac{-3 \pm \sqrt{41}}{4}$

B $x = \frac{3 \pm \sqrt{41}}{4}$

C $x = \frac{3 \pm \sqrt{23}}{4}$

D $x = \frac{-3 \pm \sqrt{23}}{4}$

E There are no real solutions

2 Write $x^2 - 12x + 3$ in completed square form.

A $(x - 6)^2 + 3$

B $(x - 12)^2 - 141$

C $(x + 6)^2 - 33$

D $(x - 6 - \sqrt{33})(x - 6 + \sqrt{33})$

E $(x - 6)^2 - 33$

3 The curve $y = -2(x - 5)^2 + 3$ meets the y-axis at A and has a maximum point at B. Find the coordinates of A and B.

A A(0, 3) and B(5, 3)

B A(0, −47) and B(5, 3)

C A(0, −47) and B(5, −6)

D A(0, 3) and B(−10, −47)

E A(0, −47) and B(−10, 3)

4 Using the coordinates of the turning point and the equation of the line of symmetry state which one of the following is the equation of this curve.

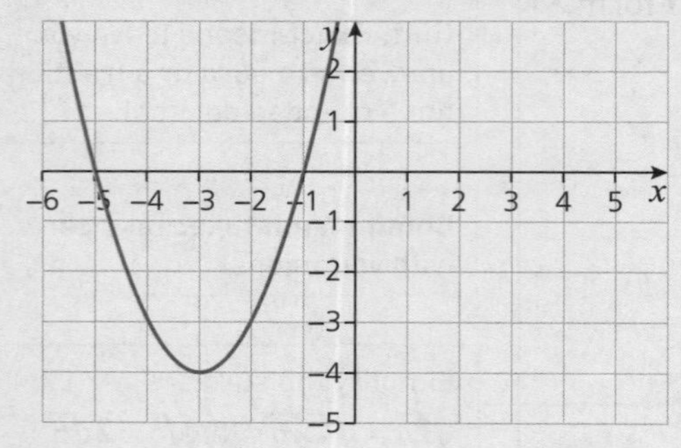

A $y = (x - 3)^2 - 4$
B $y = (x - 3)^2 + 4$
C $y = (x + 3)^2 + 4$
D $y = (x + 4)^2 - 3$
E $y = (x + 3)^2 - 4$

5 Four of the following statements are true and one is false. Which one is false?

A $3x^2 - 2x + 1 = 0$ has no real roots.
B $2x^2 - 5x + 1 = 0$ has two distinct real roots.
C $9x^2 - 6x + 1 = 0$ has one repeated real root.
D $x^2 + 2x - 5 = 0$ has two distinct real roots.
E $4x^2 - 9 = 0$ has one repeated real root.

Full worked solutions online

CHECKED ANSWERS

Exam-style question

You are given that $f(x) = 3x^2 - 12x - 6$.

i Express $f(x)$ in the form $a(x + b)^2 + c$ where a, b and c are integers.

ii The curve C with equation $y = f(x)$ meets the y-axis at P and has a minimum point at Q.

a State the coordinates of P and Q.

b Sketch the curve.

Short answers on page 152

Full worked solutions online

CHECKED ANSWERS

Chapter 4 Equations and inequalities

About this topic

You may need to solve **simultaneous equations** in many areas of mathematics including finding the coordinates of the point where two lines or a line and a curve intersect.

Inequalities are very useful for expressing ranges of values. The algebra of inequalities is very similar to the algebra of equations, but there are a couple of small, yet important, differences.

Before you start, remember

- how to solve linear and quadratic equations
- how to sketch graphs of linear and quadratic equations.

Simultaneous equations

REVISED

Key facts

1 A **linear equation** is an equation whose graph is a straight line. When you draw the graphs of two linear equations, then, unless the graphs are parallel, they intersect. The coordinates of the point of intersection give the solution to the two **linear simultaneous equations**.

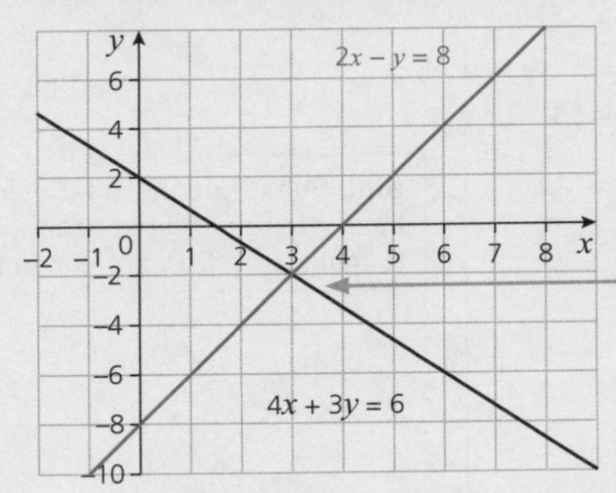

2 You should solve simultaneous equations algebraically. There are two different methods: the **elimination method** and the **substitution method**.

Linear equations do not involve terms in any powers of x or y, or terms like xy.

Examples of linear equations are $2x-3y=1$ and $y=3-4x$.

The graphs intersect at the point (3, −2) ...

... so the solution of the two linear simultaneous equations $2x-y=8$ and $4x+3y=6$ is $x=3$, $y=-2$.

Multiply one or both of the equations through by a constant, so that adding or subtracting the resulting equations eliminates one of the unknowns.

Substitute one equation into the other so the resulting equation is in terms of just one unknown.

3 **Non-linear simultaneous equations** involve terms like x^2, y^2 or other powers.
When you draw the graphs of a pair of equations where one is linear and the other quadratic then there could be different solutions as shown below.

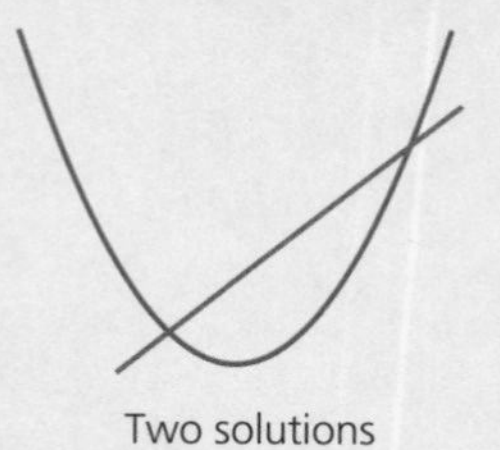

Two solutions

One repeated solution

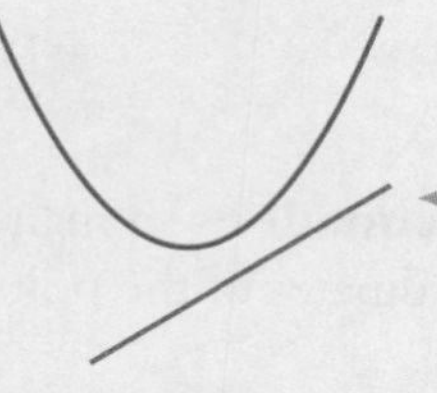

No solution

A straight line graph can cross a quadratic graph twice, touch it in one place (a tangent) or not cross it at all.

4 Use the **substitution method** to solve one linear and one quadratic equation:

- Rearrange the linear equation if necessary so that one unknown is given in terms of the other.
- Substitute the resulting equation into the quadratic equation.
- Solve to find the value(s) of one of the unknowns.
- Substitute back into the linear equation to find the values of the other unknown.

There could be 2 solutions, 1 repeated solution or 0 solutions.

Worked examples

1 Linear simultaneous equations: elimination method (1)

Solve the simultaneous equations

$$2x - y = 8 \quad ①$$
$$4x + 3y = 6 \quad ②$$

Solution

Multiply equation ① by 3: $6x - 3y = 24$

Leave equation ② as it is: $4x + 3y = 6$

Add the equations: $10x = 30 \Rightarrow x = 3$

Substitute $x = 3$ into equation ①: $6 - y = 8 \Rightarrow y = -2$

The solution is $x = 3$, $y = -2$.

Hint: Substitute the solutions back into the original equations to make sure they are correct.

2 Linear simultaneous equations: elimination method (2)

Solve the simultaneous equations

$$2x - 5y = 5 \quad ①$$
$$3x - 2y = 13 \quad ②$$

Solution

Multiply equation ① by 3: $6x - 15y = 15$

Multiply equation ② by 2: $6x - 4y = 26$

Subtract the equations: $-11y = -11 \Rightarrow y = 1$

Substitute $y = 1$ into equation ①: $2x - 5 = 5 \Rightarrow x = 5$

The solution is $x = 5$, $y = 1$.

Alternatively, you could eliminate the ys by multiplying the first equation by 2 and the second by 5.

3 Linear simultaneous equations: substitution method

Solve the simultaneous equations $3x - 4y = 2$ ①
$y = 2x + 1$ ②

Solution

Substitute equation ② into equation ①: $3x - 4(2x + 1) = 2$
$3x - 8x - 4 = 2$
$-5x = 6$
$x = -1.2$

Substitute $x = -1.2$ into equation ②: $y = -2.4 + 1 = -1.4$.

The solution is $x = -1.2$, $y = -1.4$.

Hint: The substitution method is particularly useful if one equation gives one variable (unknown) in terms of the other, or can easily be rewritten in this form.

Replace 'y' in the first equation by '$2x + 1$'.

4 One linear and one quadratic equation: substitution method

Solve the simultaneous equations $x^2 - 2y^2 = 7$ ①
$2x + y = 5$ ②

Solution

Rearrange equation ② to give y in terms of x: $y = 5 - 2x$

Substitute into equation ①: $x^2 - 2(5 - 2x)^2 = 7$

Multiply out and simplify: $x^2 - 2(25 - 20x + 4x^2) = 7$
$x^2 - 50 + 40x - 8x^2 = 7$
$-7x^2 + 40x - 57 = 0$

Multiply through by -1: $7x^2 - 40x + 57 = 0$

Factorise and solve the equation: $(7x - 19)(x - 3) = 0$
$x = \frac{19}{7}$ or 3

Substitute into equation ②: When $x = \frac{19}{7}$, $y = 5 - \frac{38}{7} = -\frac{3}{7}$
When $x = 3$, $y = 5 - 6 = -1$

The solutions are $x = \frac{19}{7}$, $y = -\frac{3}{7}$ and $x = 3$, $y = -1$.

Common mistakes: Take care with your signs!

Hint: You can factorise $7x^2 - 40x + 57$ by splitting the middle term – see Chapter 3.

Common mistakes: Don't forget to find the corresponding y-values. You must substitute back into the linear equation otherwise you will find extra false solutions.

Test yourself

TESTED

1 Which one of the following is the correct x-value for the linear simultaneous equations $5x - 3y = 1$ and $3x - 4y = 4$?

A $x = \frac{16}{29}$ B $x = \frac{16}{11}$ C $x = -\frac{8}{11}$ D $x = -\frac{3}{11}$ E $x = \frac{5}{29}$

2 Which one of the following is the correct y-value for the linear simultaneous equations $5x - 2y = 3$ and $y = 1 - 2x$?

A $y = -9$ B $y = \frac{5}{9}$ C $y = \frac{7}{9}$ D $y = -\frac{1}{9}$ E $y = -\frac{7}{3}$

3 Look at the simultaneous equations $2x^2 + y^2 = 4$ and $3x + y = 1$.
Which of the following is the quadratic equation which must be solved to find the x-values of the solution?

A $11x^2 - 6x - 3 = 0$ B $11x^2 - 3 = 0$ C $7x^2 + 6x + 3 = 0$ D $x^2 + 6x + 3 = 0$ E $11x^2 + 6x - 3 = 0$

4 Simon is solving the simultaneous equations $y(1 - x) = 1$ and $2x + y = 3$.
Simon's working is shown below.

Rearrange second equation:	$y = 2x - 3$	Line X
Substitute into first equation:	$(2x - 3)(1 - x) = 1$	Line Y
	$-2x^2 + 5x - 3 = 1$	
	$2x^2 - 5x + 4 = 0$	
Discriminant	$= -5^2 - 4 \times 2 \times 4 = -25 - 32 = -57$	Line Z
	There are no real solutions.	

Simon knows that he must have made at least one mistake, as his teacher has told him that the equations do have real solutions.
In which line(s) of the working has Simon made a mistake?

A Line X only B Line Y only C Lines Y and Z D Line Z only E Lines X and Z

5 Look at the simultaneous equations $3x^2 + 2y^2 = 5$ and $y - 2x = 1$.
Which one of the following is the correct pair of x-values for the solution of these equations?

A $x = \frac{3}{11}$ or $x = -1$ B $x = \pm\sqrt{\frac{3}{11}}$ C $x = \frac{3}{7}$ or $x = -1$ D $x = -\frac{3}{11}$ or $x = 1$ E $x = -\frac{3}{7}$ or $x = 1$

Full worked solutions online

CHECKED ANSWERS ☐

Exam-style question

The diagram is a plan view of a rectangular enclosure. A wall forms one side of the enclosure. The other three sides are formed by fencing of total length 34 m. The width of the rectangle is x m, the length is y m, and the area enclosed is 144 m².

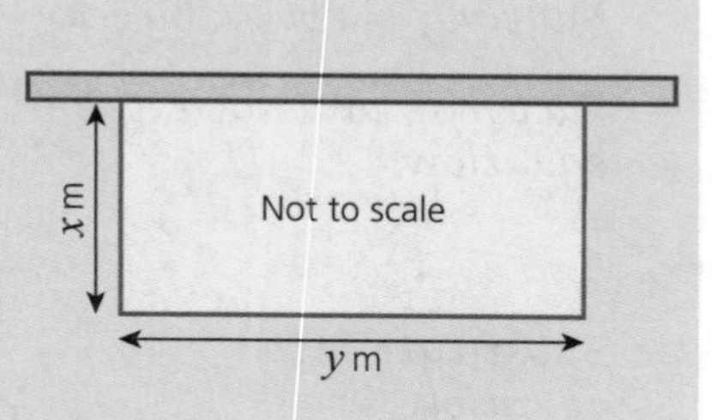

i Write down two equations involving x and y.
ii Hence show that $x^2 - 17x + 72 = 0$.
iii By factorising, solve this equation and find the possible dimensions of the rectangle.

Short answers on page 152

Full worked solutions online

CHECKED ANSWERS ☐

Inequalities

REVISED

Key facts

1 Solving linear inequalities is very similar to solving linear equations; however, you should bear in mind the following:

- When swapping the sides of an inequality you should reverse the direction of the inequality sign.
- When multiplying or dividing both sides of an inequality by a negative you should reverse the direction of the inequality. It is often best to avoid multiplying or dividing by a negative number if you can.

$6 < x$ so $x > 6$.

$-2x < 6$ so $x > -3$.

2 To solve a quadratic inequality, you should solve the corresponding quadratic equation and then use a graph to determine the solution to the inequality.

The quadratic graph has a positive y-value here. So the range of values where the quadratic is > 0 is $x < x_1$ or $x > x_2$.

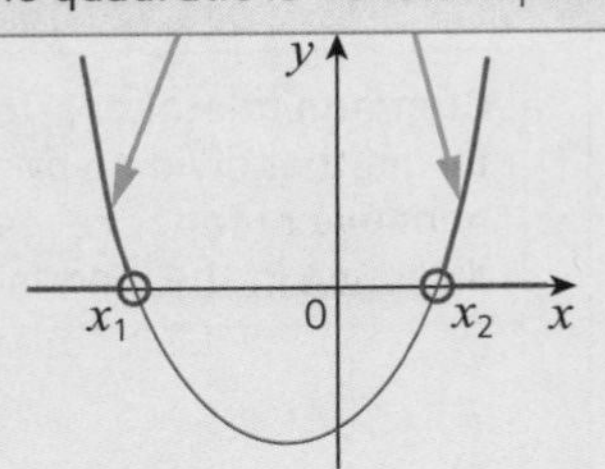

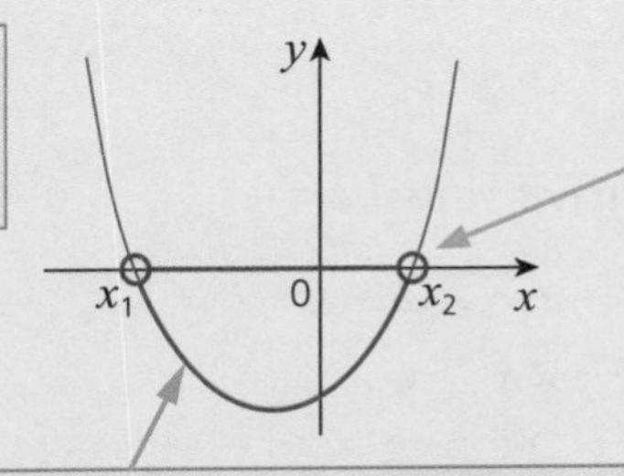

Use an open circle, ∘, to show the value of x_1 **is not** included.

Use a filled in circle, •, to show the value of x_1 **is** included.

The quadratic graph has a negative y-value here. So the range of values where the quadratic is < 0 is $x_1 < x < x_2$.

3 You can write the solutions to inequalities using set notation.

For example: $x < -2$ or $x \geqslant 5$ is $\{x : x < -2\} \cup \{x : x \geqslant 5\}$.

and $-2 \leqslant x < 5$ is $\{x : -2 \leqslant x\} \cap \{x : x < 5\}$.

x belongs to the union of the set of numbers which are less than –2 and the set of the numbers which are greater or equal to 5.

x belongs to the intersection of the set of numbers which are greater or equal to –2 and the set of the numbers which are less than 5.

4 You can represent inequalities graphically by sketching the line and then shading the relevant side of the line.

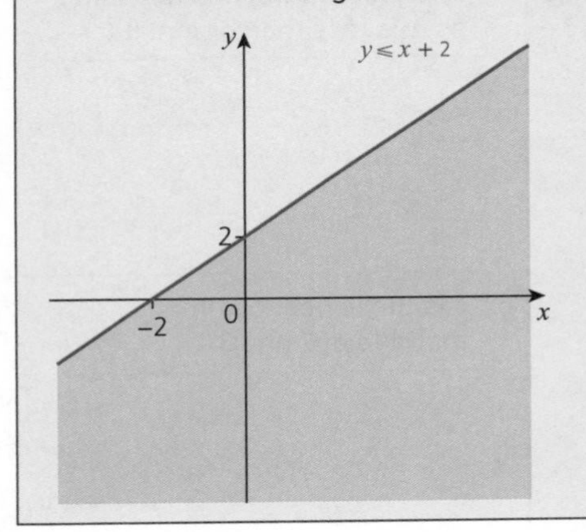

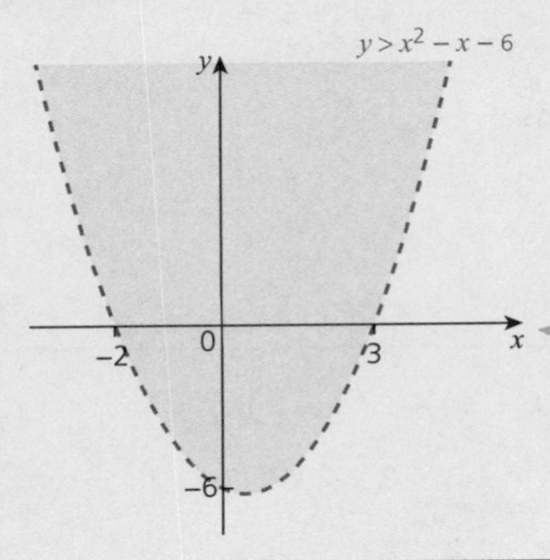

Use dashed line, - - - -, to show the line **is not** included.

Use a solid line, ———, to show the line **is** included.

Worked examples

1 Solving a linear inequality

Solve the inequality $5x - 7 < 2x + 2$.

Solution

Add 7 to both sides: $5x < 2x + 9$

Subtract $2x$ from both sides: $3x < 9$

Divide by 3: $x < 3$

The solution to an inequality looks like the solution to an equation but it has an inequality sign instead of an equals sign. x should be the subject.

2 Linear inequality – changing the direction of the inequality

Solve the inequality $4 - 3x \geqslant 6 - x$.

Solution

Method 1 – dividing by a negative number

$4 - 3x \geqslant 6 - x$

Add x to both sides: $4 - 2x \geqslant 6$

Subtract 4 from both sides: $-2x \geqslant 2$

Divide both sides by -2: $x \leqslant -1$

Common mistakes: Don't forget that dividing by a negative number reverses the direction of the inequality sign.

Method 2 – avoids dividing by a negative number

$4 - 3x \geqslant 6 - x$

Add $3x$ to both sides: $4 \geqslant 6 + 2x$

Subtract 6 from both sides: $-2 \geqslant 2x$

Divide both sides by 2: $-1 \geqslant x$

Rewrite with x the subject: $x \leqslant -1$

Hint: Inequalities should be written with x as the subject. Notice that '-1 is greater than or equal to x' means the same as 'x is less than or equal to -1'.

3 Linear inequality with fractions

Solve the inequality $\frac{x-4}{2} < \frac{4-x}{3}$.

Solution

Multiply by 6: $3(x - 4) < 2(4 - x)$

Expand: $3x - 12 < 8 - 2x$

Add $2x$ to both sides: $5x - 12 < 8$

Add 12 to both sides: $5x < 20$

Divide both sides by 5: $x < 4$

6 is the lowest common multiple of 2 and 3.

4 Solving graphical problems using inequalities

Sketch the lines $y=3x+1$ and $y=-2x-4$.

For what values of x is the line $y=3x+1$ above the line $y=-2x-4$?

Solution

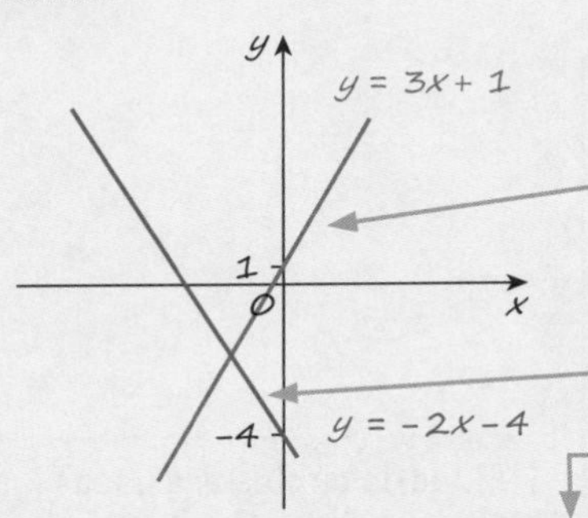

$y=3x+1$ cuts the y-axis at $y=1$ and has a gradient of 3.

$y=-2x-4$ cuts the y-axis at $y=-4$ and has a gradient of -2.

'Above' means that the y-value for one will be bigger than the y-value for the other for the same value of x.

The line $y=3x+1$ is above the line $y=-2x-4$ when $3x+1>-2x-4$.

Solve: $3x+1>-2x-4$

Add $2x$ to both sides: $5x+1>-4$

Subtract 1 from both sides: $5x>-5$

Divide both sides by 5: $x>-1$

Hint: Check that the solution looks right on the diagram. $y=3x+1$ is above $y=-2x-4$ to the right of $x=-1$.

5 Solving quadratic inequalities that use > or <

Solve the inequality $x^2+x-6>0$.

Solution

The corresponding quadratic equation is $x^2+x-6=0$.

Just treat the inequality as a quadratic equation: solve it normally and then draw the graph.

This factorises into $(x+3)(x-2)=0$.

So the graph of $y=x^2+x-6$ intercepts the x-axis at $x=-3$ and $x=2$.

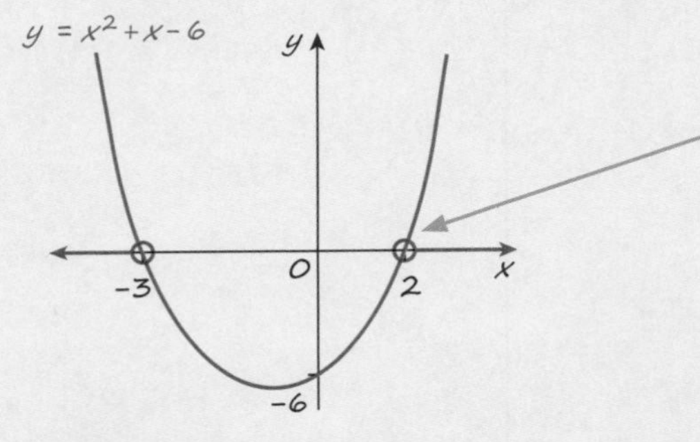

Empty circles are used to show that −3 and 2 are not included

The graph has a positive y-value (i.e. is above the x-axis) to the left of $x=-3$ and to the right of $x=2$.

$x=-3$ and $x=2$ should **not** be included, because at these points $y=0$ and the inequality you are solving is >0.

So the solution is $x<-3$ or $x>2$.

Or using set notation this is:
$\{x : x<-3\} \cup \{x : x>2\}$

Common mistakes: Two regions need two inequalities to describe them. Don't combine this into one inequality saying '$-3>x>2$'. This would mean that x was less than −3 and greater than 2 at the same time which is impossible!

Hint: Check that the solution looks right on the diagram. There are two regions on the graph where $x^2+x-6>0$.

6 Solving quadratic inequalities that use ⩽ or ⩾

Solve the inequality $x^2+2x+2 \leqslant -4x-6$.

Solution

Rearrange to make one side zero: $x^2+6x+8 \leqslant 0$

The corresponding quadratic is: $x^2+6x+8=0$

This factorises into $(x+4)(x+2)=0$

So the graph of $y=x^2+6x+8$ intercepts the x-axis at $x=-4$ and $x=-2$.

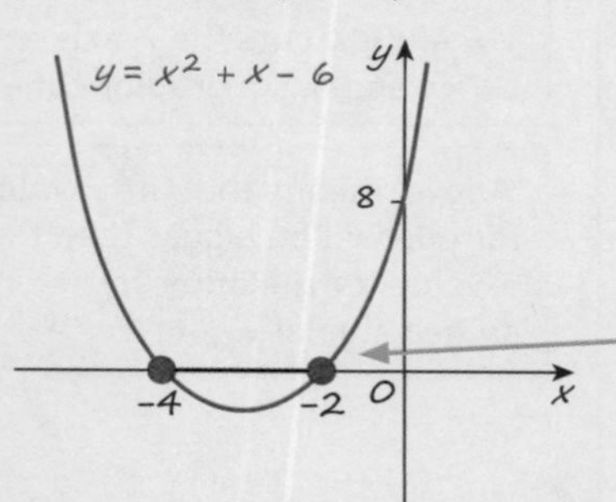

Filled-in circles are used to show that −4 and −2 are included.

The graph has a negative y-value (i.e. is below the x-axis) between $x=-4$ and $x=-2$.

$x=-4$ and $x=-2$ should be included, because at these points $y=0$ and the inequality you are solving is $\leqslant 0$.

So the solution is $-4 \leqslant x \leqslant -2$

Or using set notation this is $\{x: -4 \leqslant x\} \cap \{x: x \leqslant -2\}$

One region only needs one inequality to describe it.

Note: You could solve the above example by drawing the graphs of $y=x^2+2x+2$ and $y=-4x-6$ and finding where they cross, but it's much easier to rearrange to make a quadratic equal to zero and just draw a single graph.

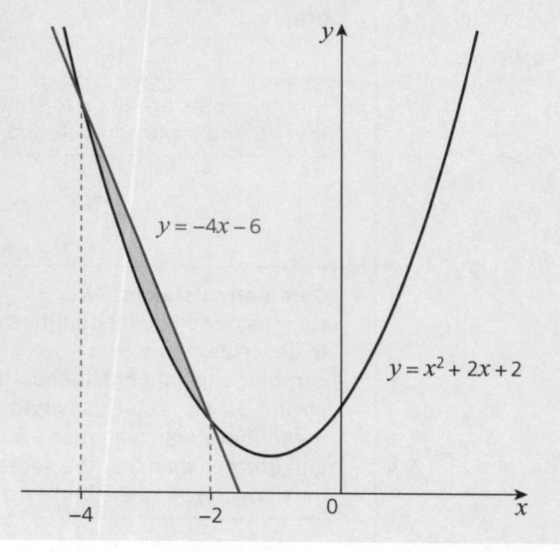

Test yourself

TESTED

1 Solve $x + 7 < 3x - 5$.

A $x > 6$ B $x > 1$ C $x < 1$ D $x < 6$ E $x > 4$

2 Solve $\frac{2(2x+1)}{3} \geqslant 6$.

A $x \geqslant 5$ B $x \geqslant \frac{3}{2}$ C $x > 4$ D $x \geqslant 4$ E $x > 5$

3 The graph shows the line $y = 3x - 3$ and $y = -x + 5$. For what values of x is $y = 3x - 3$ above $y = -x + 5$?

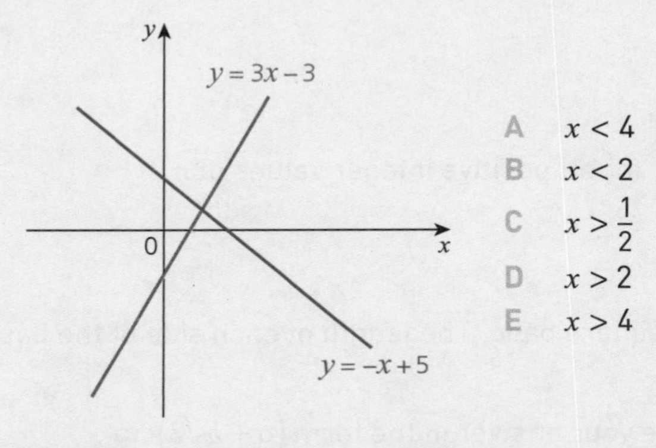

A $x < 4$

B $x < 2$

C $x > \frac{1}{2}$

D $x > 2$

E $x > 4$

4 Solve the inequality $x^2 + 2x - 15 \leqslant 0$.

A $-5 \leqslant x \leqslant 3$ B $-3 \leqslant x \leqslant 5$ C $3 \leqslant x \leqslant 5$ D $x = -5$ or $x = 3$ E $x \leqslant -5$ or $x \geqslant 3$

5 Solve the inequality $6x - 6 < x^2 - 1$.

A $x < -1$ or $x > 7$ B $1 < x < 5$ C $x < 1$ or $x > 7$ D $x < -5$ or $x > -1$ E $x < 1$ or $x > 5$

Full worked solutions online

CHECKED ANSWERS

Exam-style question

i Solve the inequality $\frac{2(3x-4)}{5} \leqslant -4$

ii Solve the inequality $x^2 - x - 2 > 4$

Short answers on page 152

Full worked solutions online

CHECKED ANSWERS

Review questions (Chapters 1–4)

1 P: n is a multiple of 2

Q: n is a multiple of 3

R: n is an even multiple of 3

S: n is a multiple of 6

Insert the correct symbol ($\Rightarrow, \Leftarrow$ or $\Leftrightarrow$) into each box.

i P ☐ S **ii** S ☐ Q **iii** R ☐ S

2 i Prove that $n^2 - n$ is even for any integer value of n.

ii Disprove the conjecture: $n^2 + 1$ is a prime number for all positive integer values of n.

3 i Factorise $2x^2 + 5x - 3$.

ii Sketch the curve $y = 2x^2 + 5x - 3$.

4 A rectangular block of volume $7(8 - 5\sqrt{2})\text{cm}^3$ has a square base. The length of each side of the base is $(3 - \sqrt{2})\text{cm}$.

i Find the area of the square base of the block. Give your answer in the form $(a + b\sqrt{2})\text{cm}^2$.

ii Prove that the height of the block is $(c - \sqrt{2})\text{cm}$ where c is a constant to be found.

5 i You are given that $2^{x+y} \times 3^y = 2^5 \times 3^{2x-1}$. Find the value of x and y.

ii Solve $x^4 - 5x^2 + 4 = 0$.

6 i Express $f(x) = x^2 - 10x + 4$ in the form $(x + p)^2 + q$ where p and q are integers.

ii The equation $2x^2 + kx + 8 = 0$ has no real roots. Find the possible values of k.

Short answers on page 152

Full worked solutions online

CHECKED ANSWERS ☐

SECTION 2

Target your revision (Chapters 5–9)

1 Calculate the length, midpoint and gradient of a line segment

Two points A and B have coordinates (2, 5) and (−6, 9) respectively.
Find:
- i the midpoint of AB
- ii the gradient of AB
- iii the distance AB.

(see page 31)

2 Use the relationship between the gradients of parallel lines and find the equation of a line

The line l passes through the point (−2, 1) and is parallel to the line $x - 2y = 5$.
- i Find the equation of the line l, giving your answer in the form $y = mx + c$.
- ii Draw a graph of the line l for $-6 \leqslant x \leqslant 4$.

(see page 35)

3 Use the relationship between the gradients of perpendicular lines

The points A and B have coordinates (2, −4) and (3, 6).
Find the gradient of a line perpendicular to AB.

(see page 35)

4 Find the coordinates of the point where two lines intersect

The lines $6x - 9y = 7$ and $y = 4x - 3$ intersect at the point X.
Find the coordinates of X.

(see page 45)

5 Use the equation of a circle

Find the equation of the following circles:
- i centre (2, −3), radius 4
- ii passing through the points A(−1, 0) and B(3, 6) with diameter AB.

(see page 40)

6 Find the coordinates of the point where two curves intersect

Find the coordinates of the point(s) where the curves $(x-2)^2 + y = 5$ and $y = 2x^2 - 3$ intersect.

(see page 45)

7 Solve problems involving circles

Find the equation of the tangent to the circle $(x-1)^2 + (y+2)^2 = 25$ at the point (5, 1).

(see page 40)

8 Use exact values of sin θ, cos θ and tan θ

Without using your calculator find the exact values of:

i $\sin 300°$ ii $\cos 120°$ iii $\tan 405°$

(see page 50)

9 Solve trigonometric equations

Solve the following equations for $0° \leqslant x \leqslant 180°$
- i $\sin x = \frac{\sqrt{3}}{2}$
- ii $\sin 3x = \frac{\sqrt{3}}{2}$
- iii $\sin(x + 45°) = \frac{\sqrt{3}}{2}$

(see page 50)

10 Use trigonometric identities

Prove that $\tan\theta + \frac{1}{\tan\theta} \equiv \frac{1}{\sin\theta\cos\theta}$.

(see page 50)

11 Solve problems involving triangles without right angles

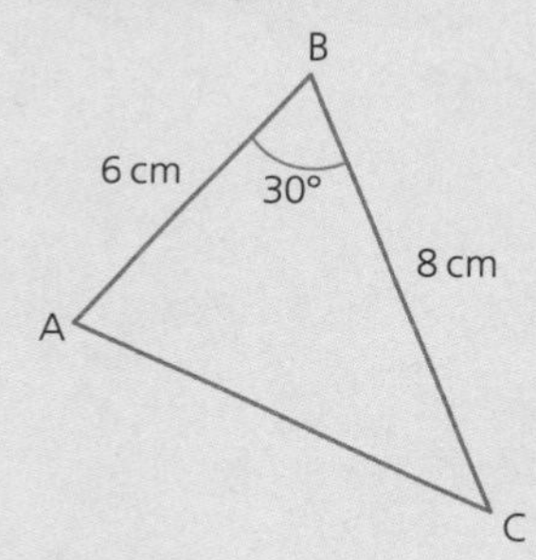

Find:
- i the area of triangle ABC
- ii the length AC.

(see page 55)

12 Add, subtract, multiply and divide polynomials

You are given $f(x) = 5x^3 - 2x + 3$, $g(x) = 2x^2 - 3x - 2$ and $h(x) = x + 1$.
Find:

i $f(x) + g(x)$ ii $f(x) - g(x)$

iii $f(x) \times g(x)$ iv $f(x) \div h(x)$.

(see pages 60 and 65)

13 Use the factor theorem and sketch the graph of a polynomial

You are given $f(x) = 2x^3 + x^2 - 5x + 2$.

i Show that $x = 1$ is a root of $f(x) = 0$.

ii Show that $(x + 2)$ is a factor of $f(x)$.

iii Factorise $f(x)$ fully and hence solve $f(x) = 0$.

iv Sketch the graph of $y = f(x)$.

(see page 68)

14 Use direct and inverse proportion

y is inversely proportional to the cube of x.

When $x = 2$, $y = 2$.

Find the equation connecting x and y.

(see page 72)

15 Sketch the graph of trigonometric functions by using transformations

For $0° \leqslant x \leqslant 360°$, sketch the graphs of:

i $y = \tan x$

ii $y = -\tan x$

iii $y = \cos \frac{1}{2}x$

iv $y = 1 + \sin x$.

(see page 78)

16 Use the binomial expansion

i Expand $(2 - 3x)^4$.

ii Find the coefficient of x^7 in the expansion of $(2 - 3x)^{10}$.

(see page 82)

Short answers on page 152–3

Full worked solutions online

CHECKED ANSWERS

Chapter 5 Coordinate geometry

About this topic

This chapter covers the geometry of straight lines and circles. You can use coordinate geometry to solve problems involving points, lines and circles.

Before you start, remember

- how to use coordinates from GCSE
- how to solve linear and quadratic equations
- how to solve simultaneous equations
- Pythagoras' theorem.

Working with coordinates

REVISED

Key facts

1 The diagram shows the line joining $A(x_1, y_1)$ and $B(x_2, y_2)$

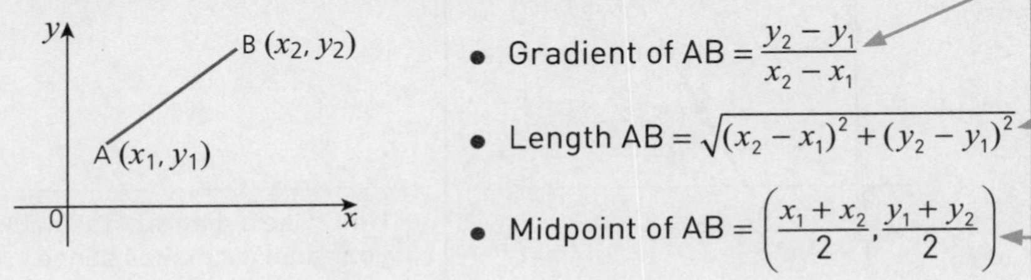

- Gradient of AB $= \dfrac{y_2 - y_1}{x_2 - x_1}$
- Length AB $= \sqrt{(x_2 - x_1)^2 + (y_2 - y_1)^2}$
- Midpoint of AB $= \left(\dfrac{x_1 + x_2}{2}, \dfrac{y_1 + y_2}{2}\right)$

This is $\dfrac{\text{difference in } y\text{'s}}{\text{difference in } x\text{'s}}$.

Length of a line is just Pythagoras.

Find the mean of the x coordinates and the mean of the y coordinates.

2 Parallel lines have the same gradient, $m_1 = m_2$

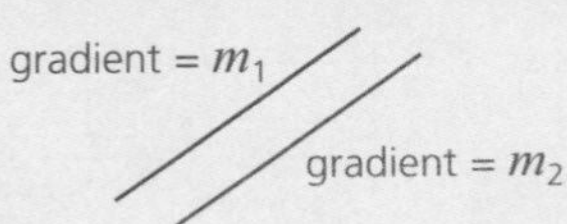

3 Perpendicular lines have gradients such that $m_1 m_2 = -1$.

This is sometimes written as: $m_2 = \dfrac{-1}{m_1}$.

gradient = m_1

gradient = m_2

Worked examples

1 The gradient of a line

For the points A(5, 2) and B(7, 8) find the gradient of the line AB and the gradient of the line perpendicular to AB.

> **Hint:** It helps to draw a diagram, even if the question does not ask for it.

> **Common mistake**: It doesn't matter which point you choose to be (x_1, y_1) or (x_2, y_2) but you must be consistent!

Solution

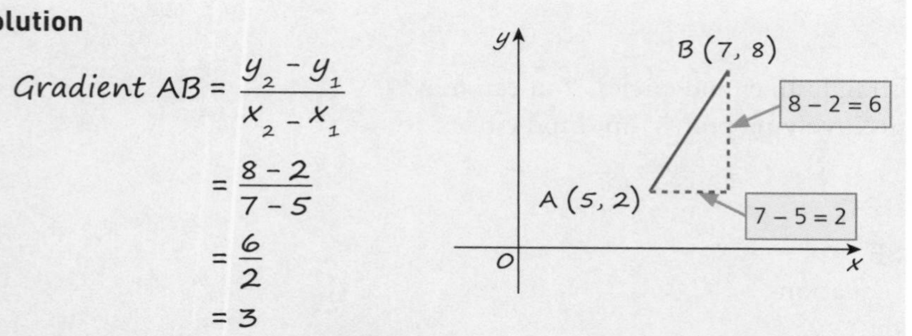

$$\text{Gradient AB} = \frac{y_2 - y_1}{x_2 - x_1}$$

$$= \frac{8-2}{7-5}$$

$$= \frac{6}{2}$$

$$= 3$$

$$\text{Gradient of perpendicular} = -\frac{1}{3}$$

> This is the negative reciprocal of 3.

2 The midpoint of a line

Find the midpoint of A(3, −1) and B(7, 6).

Solution

$$\text{Midpoint, M} = \left(\frac{x_1 + x_2}{2}, \frac{y_1 + y_2}{2}\right)$$

$$= \left(\frac{3+7}{2}, \frac{(-1)+6}{2}\right)$$

$$= (5, 2.5)$$

> **Hint:** Use a diagram to check your answer makes sense.

3 The length of the line joining two points

What is the length of the line joining A(−1, 6) and B(6, 2)?

Solution

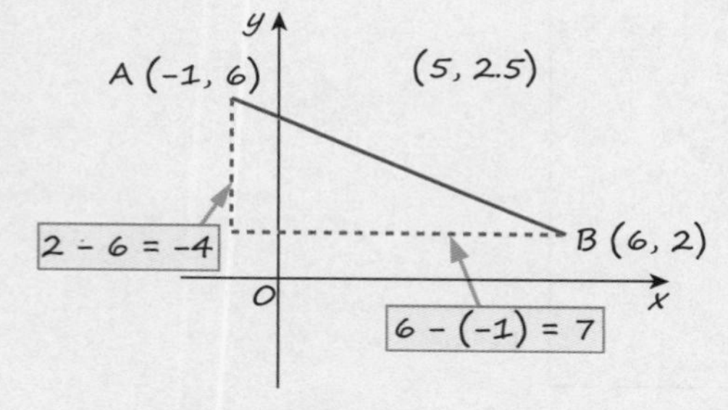

$$\text{Length AB} = \sqrt{(x_2 - x_1)^2 + (y_2 - y_1)^2}$$

$$= \sqrt{(6-(-1))^2 + (2-6)^2}$$

$$= \sqrt{(7)^2 + (-4)^2}$$

$$= \sqrt{49 + 16}$$

$$= \sqrt{65}$$

> **Common mistake**: Make sure you use brackets to avoid errors with negative signs.

4 Solving problems (1)

Three points A, B and C have coordinates (–1, 6), (1, 2) and (3, y).

AB is perpendicular to BC. Find the value of y.

Solution

$$\text{Gradient AB} = \frac{2-6}{1-(-1)} = \frac{-4}{2} = -2$$

$$\Rightarrow \text{Gradient BC} = \frac{-1}{-2} = \frac{1}{2}$$

$$\frac{y-2}{3-1} = \frac{1}{2}$$

$$\Rightarrow \frac{y-2}{2} = \frac{1}{2}$$

$$\Rightarrow y-2 = 1$$

$$\Rightarrow y = 3$$

The gradient is the negative reciprocal of –2.

Write out the formula for the gradient of the line joining B and C (with y in it).

Gradient is $\frac{\text{difference in } y\text{'s}}{\text{difference in } x\text{'s}}$.

5 Solving problems (2)

A quadrilateral has vertices A(1, –2), B(4, 0), C(2, 3) and D(–1, 1).

i Draw the quadrilateral ABCD.

ii Find the gradients of lines AB, BC, CD and DA.

iii Find the lengths of AB, BC, CD and DA.

iv What do **ii** and **iii** tell you about ABCD?

v What is the area of ABCD?

Solution

i

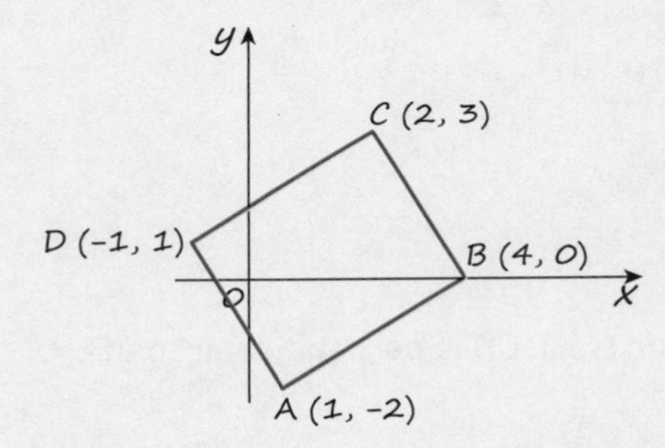

ii $\text{Gradient AB} = \frac{0-(-2)}{4-1} = \frac{2}{3}$

$\text{Gradient BC} = \frac{3-0}{2-4} = -\frac{3}{2}$

$\text{Gradient CD} = \frac{3-1}{2-(-1)} = \frac{2}{3}$

$\text{Gradient DA} = \frac{1-(-2)}{-1-1} = -\frac{3}{2}$

The gradients of BC and CD are negative reciprocals of each other, so the sides are at right angles.

The gradients of BC and DA are the same, so the sides are parallel.

Chapter 5 Coordinate geometry

iii Length AB = $\sqrt{(4-1)^2 + (0-(-2)^2}$

$= \sqrt{3^2 + 2^2}$

$= \sqrt{9+4}$

$= \sqrt{13}$

Length BC = $\sqrt{13}$

Length CD = $\sqrt{13}$

Length DA = $\sqrt{13}$

iv ABCD is a square

v $\sqrt{13} \times \sqrt{13} = 13$ square units

All the lines are the same length.

All sides are the same length and adjacent sides are perpendicular.

Test yourself

TESTED

1 The points A and B have coordinates (–1, 4) and (3, –2).
What is the gradient of the line joining A to B?

A –3 B $-\frac{2}{3}$ C –1 D –1.5 E 1.5

2 The midpoint of the line AB is (–2, 1). The coordinates of point B are (1, –1).
What are the coordinates of A?

A (–5, 3) B (–3, 3) C (–0.5, 0) D (4, –3) E (–5, 1)

3 The points A and B have coordinates (–3, 1) and (2, 5).
Find the length of the line AB.

A 3 B $\sqrt{41}$ C $\sqrt{17}$ D 41 E 5

4 The trapezium PQRS has vertices P(0, 1), Q(6, 4), R(4, z) and S(0, 5).
What is the value of z?

A 13
B 7
C 8
D $4\frac{1}{3}$
E 5

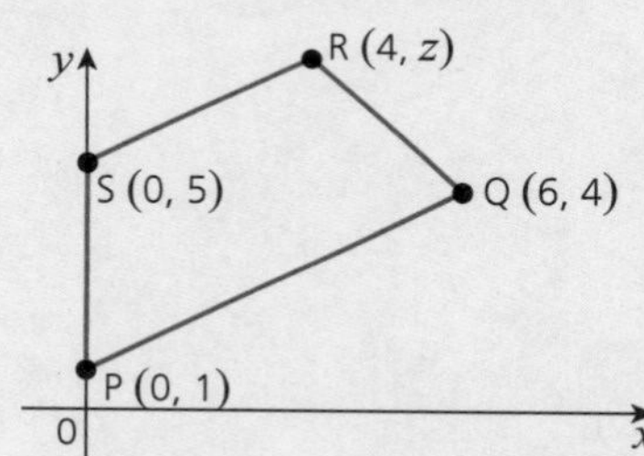

5 The points A, B, C and D have coordinates (1, 5), (3, –1), (1, 2) and (x, y). CD is perpendicular to AB.
Which of the following could be the coordinates of D?

A (7, 4) B (2, –1) C (2, 5) D (3, 8) E (7, 0)

Full worked solutions online

CHECKED ANSWERS

Exam-style question

A(2, 7), B(6, –1) and C(0, 1) are three points. M is the midpoint of AB.

i Find the coordinates of M.

ii Find the exact values of the lengths of the lines AB and CM.

iii Show that AB and CM are perpendicular and hence find the area of the triangle ABC.

Short answers on page 153

Full worked solutions online

CHECKED ANSWERS

The equation of a straight line

REVISED

1 In the equation of the line $y = mx + c$:

m is the gradient

c is the intercept with the y-axis.

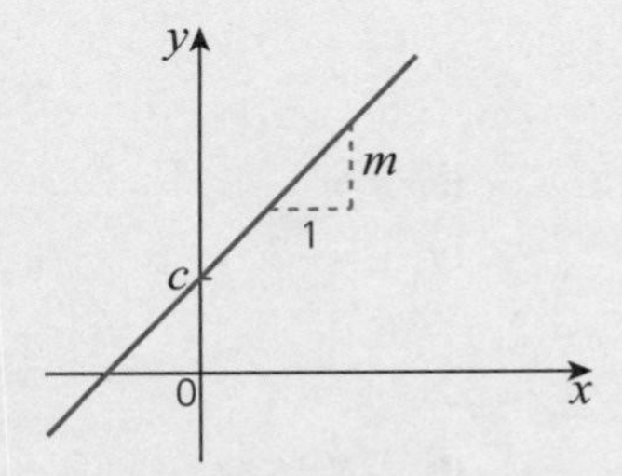

2 The equation of the line with gradient m passing through the point (x_1, y_1) is

$$y - y_1 = m(x - x_1).$$

3 The equation of the line passing through the points (x_1, y_1) and (x_2, y_2) is

$$\frac{y - y_1}{y_2 - y_1} = \frac{x - x_1}{x_2 - x_1}.$$

4 The equation of a line can also be written in the form $ax + by + c = 0$.

Usually this is written so that a, b and c are integers.

5 Vertical lines have equation $x = a$.

These lines are parallel to the y-axis.

6 Horizontal lines have equation $y = b$.

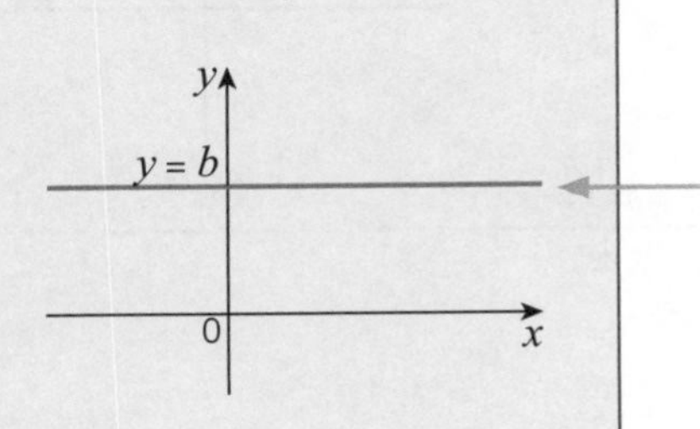

These lines are parallel to the x-axis.

7 There are several different ways to find the equation of a line.

- By:
  - obtaining its gradient,
  - substituting this value in for m in $\boldsymbol{y = mx + c}$
  - and then finding c by substituting a point into the resulting equation.
- If the gradient, m, of the line and a point (x_1, y_1) are known the equation of the line can be found with
  $\boldsymbol{y - y_1 = m(x - x_1)}$.
- If two points on the line are known the equation can be found with
  $$\frac{y - y_1}{y_2 - y_1} = \frac{x - x_1}{x_2 - x_1}.$$

Worked examples

1 Sketching lines

Sketch the following lines:

i $y = 3x + 1$ ii $x = 1$

iii $3x + 2y = 12$ iv $y = -3$.

Hint: Note that **sketching** is different from **plotting**. A sketch should just show the general shape of the graph and any important points.

Always use a ruler to draw a straight line.

Solution

i $y = 3x + 1$

Gradient 3 means 'up 3' every time you go along 1.

iii 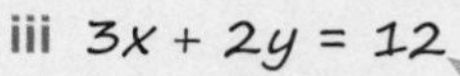

A quick way to sketch this is to find the intercepts with the axes ...

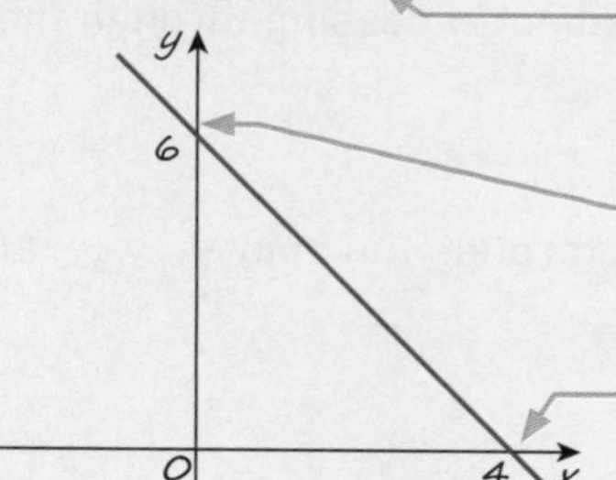

... when $x = 0$, $2y = 12$ so $y = 6$.

... and when $y = 0$, $3x = 12$ so $x = 4$.

ii 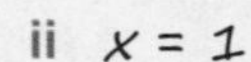

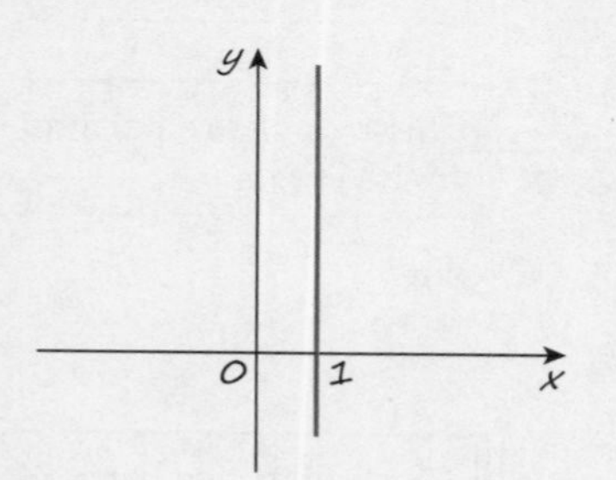

iv $y = -3$

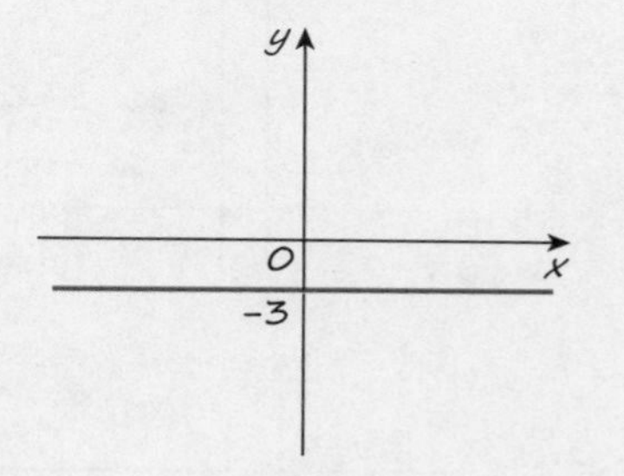

2 Finding the equation of a line parallel to another line

Find the equation of the line parallel to $y = 3x + 2$ through $(1, -5)$.

Solution

Method 1:

The gradient of $y = 3x + 2$ is 3.

Equation is: $y = 3x + c$

Substituting $(1, -5)$: $-5 = 3 \times 1 + c$

$-5 = 3 + c$ so $c = -8$

Equation: $y = 3x - 8$

Parallel lines have the same gradient.

Method 2:

Equation: $y - (-5) = 3(x - 1)$

$y + 5 = 3x - 3$

$y = 3x - 8$

Using $y - y_1 = m(x - x_1)$ where $m = 3$, $x_1 = 1$ and $y_1 = -5$.

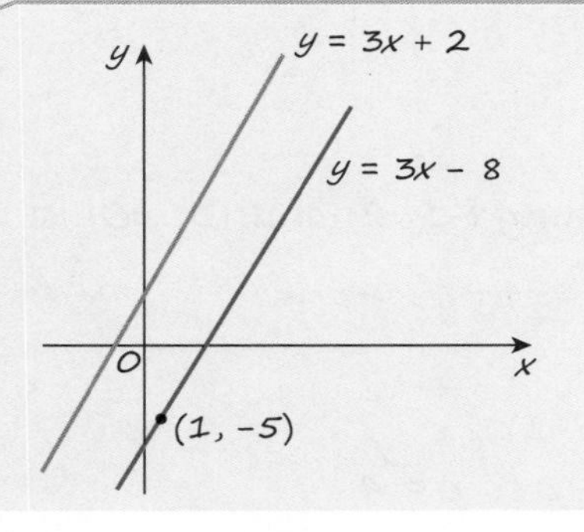

3 Finding the equation of a line perpendicular to another line

Find the equation of the line perpendicular to $y = -2x + 1$ through $(4, 0)$.

Solution

The gradient of $y = -2x + 1$ is -2

So the gradient of the perpendicular is $\frac{-1}{-2} = \frac{1}{2}$

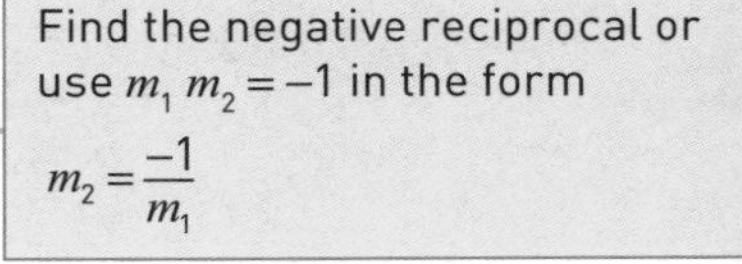

Find the negative reciprocal or use $m_1 m_2 = -1$ in the form $m_2 = \frac{-1}{m_1}$

Method 1:

Equation is: $y = \frac{1}{2}x + c$

Using $(4, 0)$: $0 = \frac{1}{2} \times 4 + c$

$c = -2$

Using $y = mx + c$.

So the equation is: $y = \frac{1}{2}x - 2$

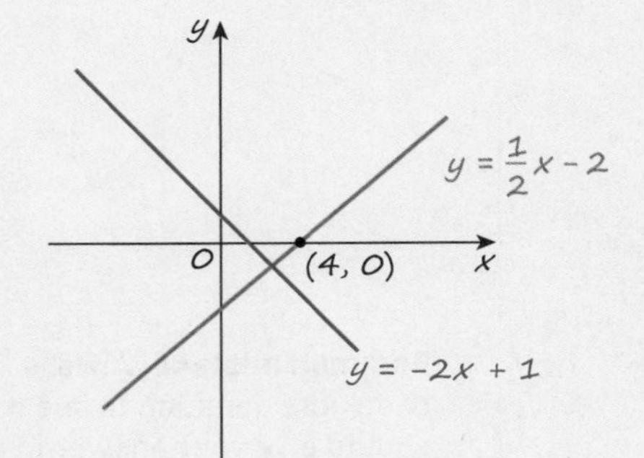

Method 2:

Equation: $y - 0 = \frac{1}{2}(x - 4)$

$y = \frac{1}{2}x - 2$

Using $y - y_1 = m(x - x_1)$ where $m = \frac{1}{2}$, $x_1 = 4$ and $y_1 = 0$.

4 Finding the equation of a line given two points

Find the equation of the line through (–1, 2) and (3, 10).

Solution

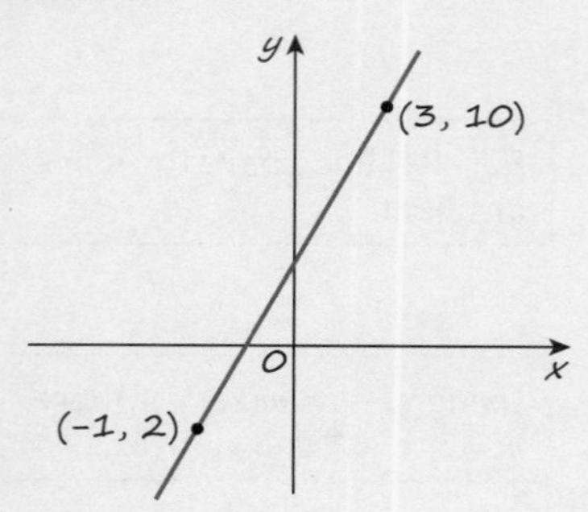

Method 1:

The gradient of the line joining (–1, 2) and (3, 10) is:

$\frac{10-2}{3-(-1)} = \frac{8}{4} = 2$

Equation is: $y = 2x + c$

Using (–1, 2): $2 = 2 \times (-1) + c$

$2 = -2 + c$ so $c = 4$

So the equation is: $y = 2x + 4$

Method 2:

Equation is: $\frac{y-2}{10-2} = \frac{x-(-1)}{3-(-1)}$

$4(y-2) = 8(x+1)$

$4y - 8 = 8x + 8$

$4y = 8x + 16$

$y = 2x + 4$

> $\frac{\text{difference in } y\text{'s}}{\text{difference in } x\text{'s}}$

> Using $y = mx + c$.

> Using $\frac{y-y_1}{y_2-y_1} = \frac{x-x_1}{x_2-x_1}$

5 Using a straight line model

The freezing point of water is 0°C or 32°F.

The boiling point of water is 100°C or 212°F.

Donna draws the conversion graph of degrees Fahrenheit, F, against degrees Celsius, C.

Find the equation of Donna's straight line, give your answer in the form $aF + bC = k$, where a, b and k are integers.

Solution

Donna plots the points (0, 32) and (100, 212).

So the gradient of her line is $\frac{212-32}{100-0} = \frac{180}{100} = \frac{9}{5}$

The intercept on the vertical axis is at (0, 32) so the equation is $F = \frac{9}{5}C + 32$

Multiply by 5: $5F = 9C + 160$

So the equation is: $5F - 9C = 160$

> **Common mistake:** Always check the question to see if you should give your answer in a particular form. You may lose marks if you don't!

Test yourself

TESTED

1 Find the equation of the line through (–1, 3) and (2, –3).

A $y = -6x - 3$ B $y = -2x + 5$ C $y = -2x + 1$ D $y = 2x + 5$ E $y = -6x + 17$

2 A line has equation $5x - 7y + 2 = 0$. Find its gradient.

A $-\frac{5}{7}$ B 5 C $\frac{5}{7}$ D $-\frac{7}{5}$ E $\frac{7}{5}$

3 Which of these is the equation of the line in the diagram?

A $x + 2y = 6$
B $y = -2x + 3$
C $2x + y = 6$
D $x + 2y = 3$
E $2y - x = 6$

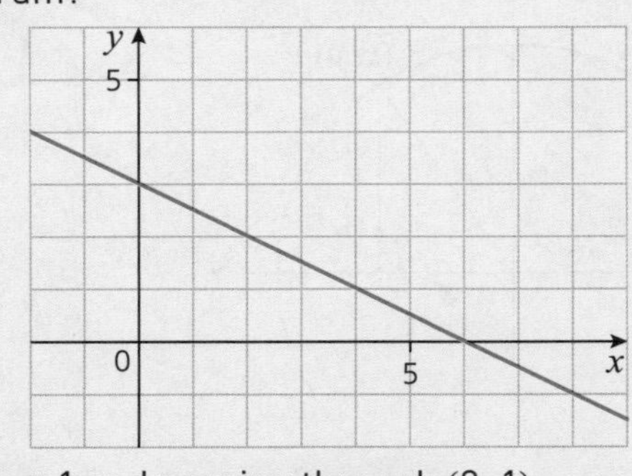

4 Find the equation of the line perpendicular to $y = -4x + 1$ and passing through (2, 1).

A $y = -\frac{1}{4}x + \frac{3}{2}$ B $y = \frac{1}{4}x + \frac{3}{2}$ C $y = \frac{1}{4}x + \frac{1}{2}$ D $y = 4x - 7$ E $y = \frac{1}{4}x + \frac{7}{4}$

5 The line L is parallel to $y = 3x - 2$ and passes through the point (–2, –1). Find the coordinates of the point of intersection with the x-axis.

A $\left(-\frac{1}{3}, 0\right)$ B $\left(-\frac{5}{3}, 0\right)$ C (0, 5) D $\left(\frac{7}{3}, 0\right)$ E $\left(\frac{2}{3}, 0\right)$

Full worked solutions online

CHECKED ANSWERS

Exam-style question

A and B are points with coordinates (–1, –2) and (3, 10) respectively.

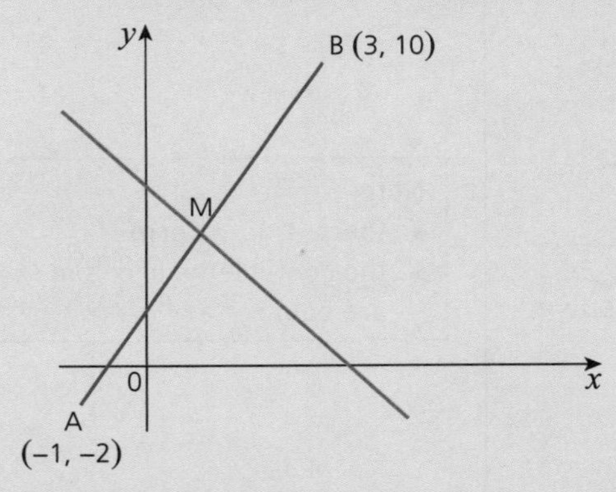

i Find the coordinates of the midpoint, M, of AB.
Show also that the equation of the perpendicular bisector of AB is $x + 3y = 13$.

ii Find the area of the triangle bounded by the perpendicular bisector, the y-axis and the line AM.

Short answers on page 153

Full worked solutions online

CHECKED ANSWERS

The circle

REVISED

Key facts

1 The equation of a circle with radius r and centre at the origin is:
$x^2 + y^2 = r^2$

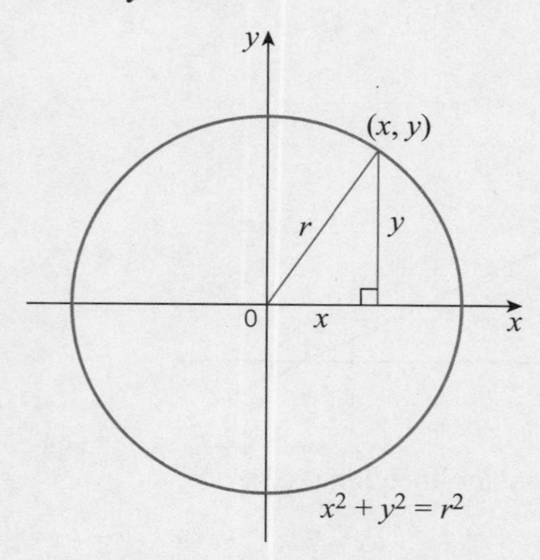

2 The equation of a circle with radius r and centre at (a, b) is:
$(x - a)^2 + (y - b)^2 = r^2$

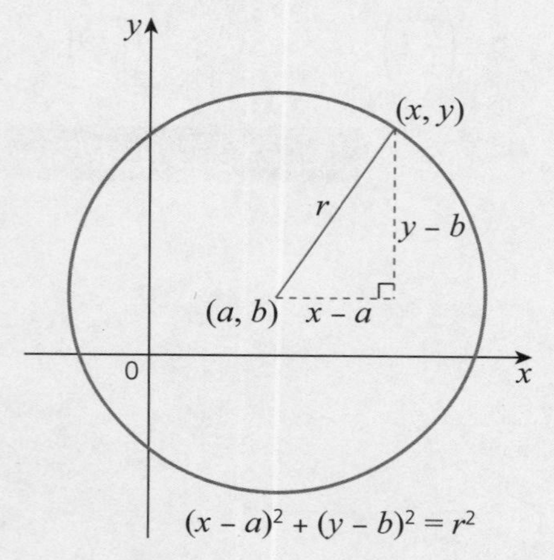

3 The equation of the circle can be rearranged and written in the form:
$x^2 + y^2 - 2ax - 2by + (a^2 + b^2 - r^2) = 0$

Note:
- there is no xy term
- the coefficients of x^2 and y^2 are equal.

4 You need to know the following **circle theorems** to help you solve problems involving circles.
- The angle in a semi-circle is a right angle.

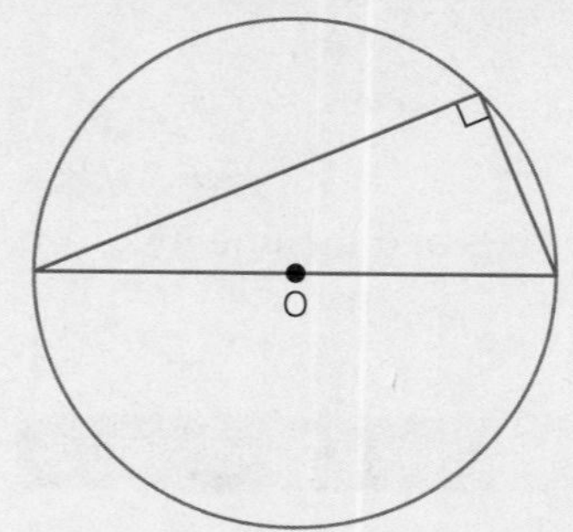

- The perpendicular from the centre of a circle to a chord bisects the chord.

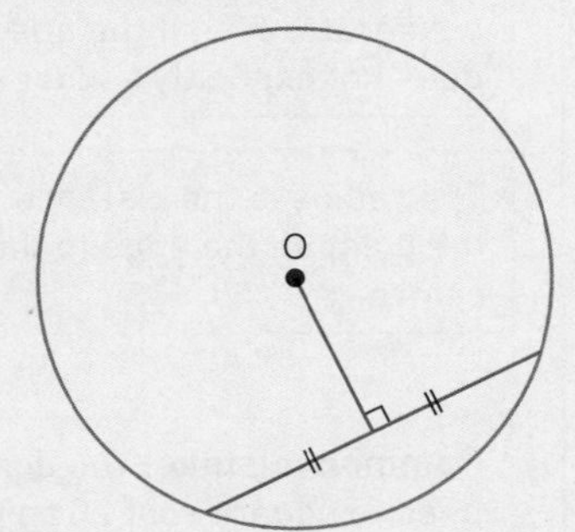

- The tangent to a circle at a point is perpendicular to the radius through that point.

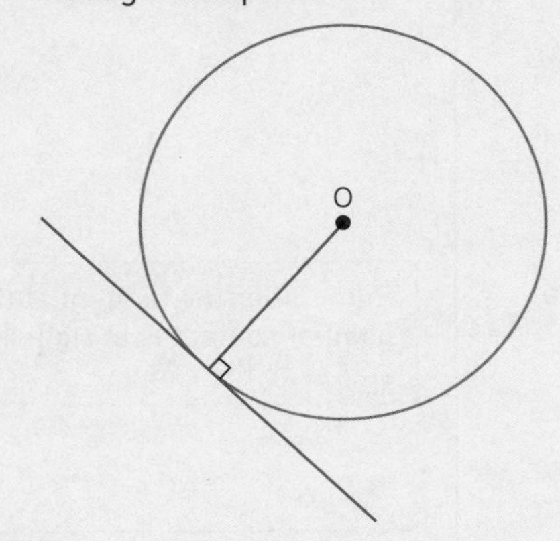

Worked examples

1 Finding the centre and radius of a circle given its equation

Give the centre and radius of the circle with equation:

$(x + 2)^2 + (y - 4)^2 = 16.$

$(x + 2)^2$ is the same as $(x - (-2))^2$ so the x coordinate of the centre is -2.

Solution

The centre is at $(-2, 4)$.

$r^2 = 16 \Rightarrow r = 4$, so the radius is 4.

2 Working with the form $x^2 + y^2 - 2ax - 2by + (a^2 + b^2 - r^2) = 0$

Show that the circle with equation $x^2 + y^2 - 8x + 2y - 8 = 0$ has centre (4, 1) and radius 5.

Solution

$$x^2 + y^2 - 8x + 2y - 8 = 0$$

$$\Rightarrow \quad x^2 - 8x + y^2 + 2y - 8 = 0$$

Collect the x and y terms.

$$\Rightarrow (x - 4)^2 - 16 + (y + 1)^2 - 1 - 8 = 0$$

Complete the square on both x and y.

$$\Rightarrow \quad (x - 4)^2 + (y + 1)^2 = 25$$

Collect the number-terms on the right-hand side.

$$\Rightarrow \quad (x - 4)^2 + (y + 1)^2 = 5^2$$

Therefore the circle has centre $(4, -1)$ and radius 5.

Common mistake: Don't write that the radius is ± 5 as the radius is a length and so is positive.

Chapter 5 Coordinate geometry

3 Finding the equation of a circle given the centre and a point on the circle

A circle has centre (3, 0) and the point (–1, 2) lies on the circumference of the circle.
Find the equation of the circle.

Hint: It is often useful to draw a diagram, even if the question does not explicitly ask for it.

Solution

$r^2 = (-1 - 3)^2 + (2 - 0)^2$

$r^2 = 4^2 + 2^2$

$r^2 = 20$

Therefore the circle has equation:

$(x - 3)^2 + y^2 = 20$

This is the same as $(y - 0)^2$.

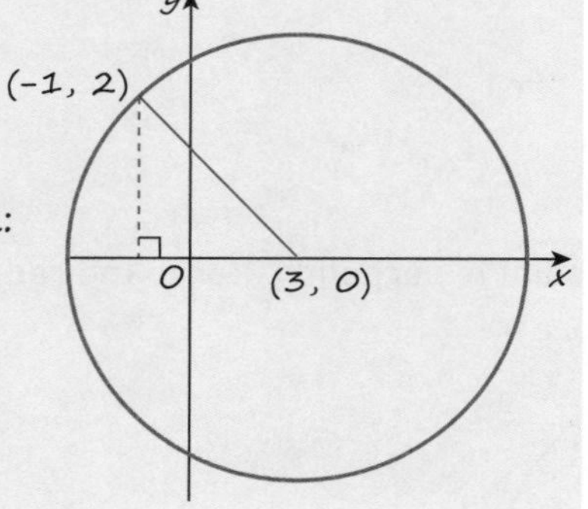

The radius is the distance from the point on the edge to the centre.

Common mistake: You don't need to square root 20 as the equation of a circle has r^2 in it.

4 Finding the equation of the tangent to the circle

Find the equation of the tangent to the circle $(x - 1)^2 + (y + 3)^2 = 10$ at the point (4, –2).

Remember the tangent at the point of contact is at right-angles to the radius.

Solution

The gradient of the radius joining the centre (1, –3) to (4, –2) is:

$$\frac{y_2 - y_1}{x_2 - x_1} = \frac{(-2) - (-3)}{4 - 1} = \frac{1}{3}$$

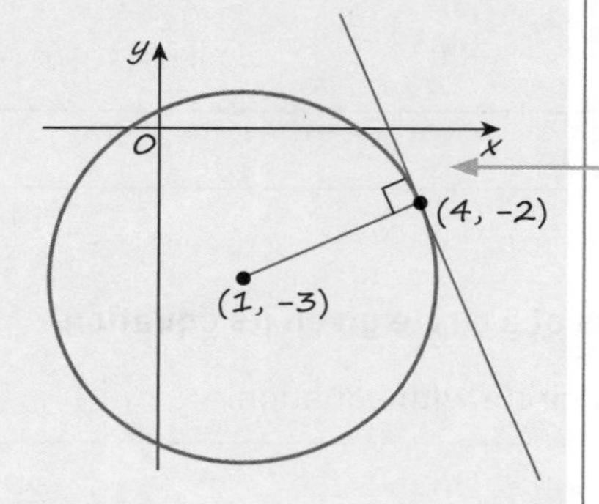

The tangent is at right-angles to the radius, so start by finding the gradient of the radius.

So the gradient of the tangent is –3

Remember $m_1m_2 = -1$ for lines at right-angles, so their gradients are negative reciprocals of each other.

The negative reciprocal of $\frac{1}{3}$ is –3.

So the equation of the tangent is

$y - (-2) = -3(x - 4)$

$y + 2 = -3x + 12$

$y = -3x + 10$

Use $y - y_1 = m(x - x_1)$ to find the equation of the line through (4, –2) with gradient –3.

5 Finding the equation of a circle given a diameter of the circle

A(1, 2), B(4, 1) and C(7, 10) are three points.

- i Show that AB and BC are perpendicular.
- ii Find the equation of the circle with AC as diameter and show that B lies on this circle.
- iii Find the coordinates of D such that BD is a diameter.

Solution

i Gradient of AB $= \frac{1-2}{4-1} = -\frac{1}{3}$

Gradient of BC $= \frac{10-1}{7-4} = 3$

$3 \times \left(-\frac{1}{3}\right) = -1$

therefore AB and BC are perpendicular.

> Using $m_1 m_2 = -1$ for perpendicular lines.

> Notice that angle ABC is therefore 90° which means B must be a point on the circle with diameter AC.

ii Midpoint, M, of AC is at $\left(\frac{1+7}{2}, \frac{2+10}{2}\right) = (4, 6)$.

> The centre of the circle is at the midpoint of AC because AC is a diameter.

AM is the radius, r, of circle

$r^2 = (4-1)^2 + (6-2)^2$

$= 3^2 + 4^2$

$= 25$

> $r^2 = AM^2 = (x_2 - x_1)^2 + (y_2 - y_1)^2$.

> The radius of the circle is $\sqrt{25} = 5$.

Equation of the circle is $(x-4)^2 + (y-6)^2 = 25$

At B: $(4-4)^2 + (1-6)^2 = 25$, therefore B is on the circle.

> **Common mistake:** Don't forget this part of the question, it's easy to miss.

iii M(4, 6) is also the midpoint of B(4, 1) and D so the coordinates of D are (4, 11).

> 6 is half-way between 1 and 11.

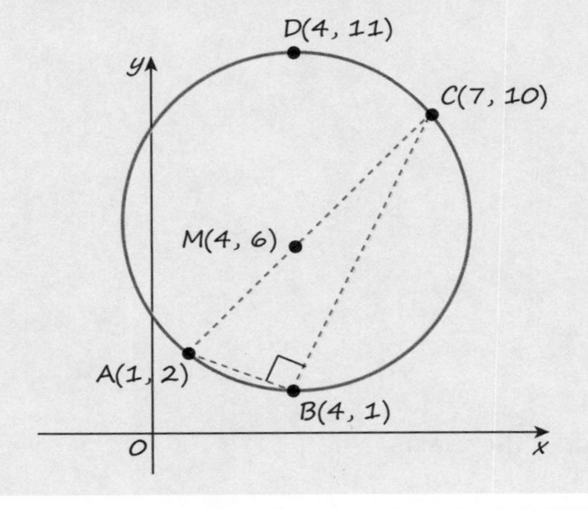

Test yourself

TESTED

1 What is the equation of the circle with centre (1, –3) and radius 5?

A $(x+1)^2+(y-3)^2=25$ B $(x-1)^2+(y+3)^2=25$ C $(x+1)^2+(y-3)^2=5$

D $(x-1)^2+(y+3)^2=5$ E $x^2+y^2=25$

2 Give the centre and the radius of the circle with equation:
$x^2+y^2+6x-4y-36=0$

A Centre (–3, 2), radius 7 B Centre (–6, 4), radius 6 C Centre (–3, 2), radius 6

D Centre (3, –2), radius 7 E Centre (3, –2), radius 6

3 Find where the circle $(x-1)^2+(y+2)^2=16$ crosses the positive x-axis.

A $x=1+4\sqrt{3}$ B $x=3$ C $x=1+2\sqrt{3}$

D $x=5$ E $x=\sqrt{15}-2$

4 Find the equation of the tangent to the circle $(x-1)^2+(y+1)^2=34$ at the point (6, 2).

A $y=-7x+44$ B $y=-\frac{5}{3}x+12$ C $y=-\frac{5}{3}x+9\frac{1}{3}$

D $y=\frac{3}{5}x-1\frac{3}{5}$ E $y=-5x+32$

5 A(2, –1) and B(4, 3) are two points on a circle with centre (1, 2). What is the distance of the chord from the centre of the circle?

A $2\sqrt{5}$ B 5 C $\sqrt{10}$

D $2\sqrt{2}$ E $\sqrt{5}$

Full worked solutions online

CHECKED ANSWERS

Exam-style question

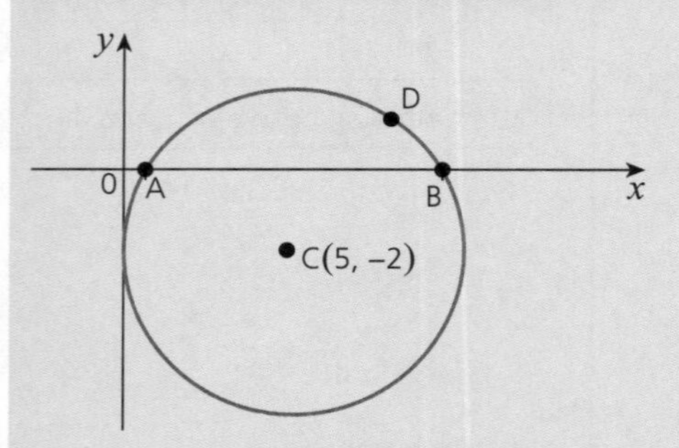

The figure shows a circle with centre C(5, –2) and radius 5.

i Show that the equation of the circle may be written as
$x^2+y^2-10x+4y+4=0$.

ii Find the coordinates of the points A and B where the circle cuts the x-axis.
Leave your answers in surd form.

iii Verify that the point D(8, 2) lies on the circle.
Find the equation of the tangent to the circle at D in the form $y=mx+c$.

Short answers on page 153

Full worked solutions online

CHECKED ANSWERS

Intersections

REVISED

Key facts

1 To find the **intersection of two lines** you need to solve a pair of linear simultaneous equations.

i.e. solve $y = m_1x + c_1$ and $y = m_2x + c_2$ simultaneously.

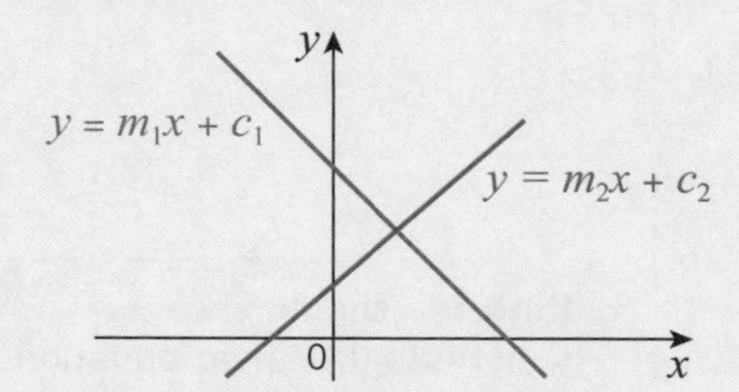

2 To find the **intersection of a curve and a line** you should solve the simultaneous equations by substituting the equation of the line into the equation of the curve.

3 A line and a curve can have more than one point of intersection, one point of intersection (which will be a tangent) or no points of intersection as shown in the figures below.

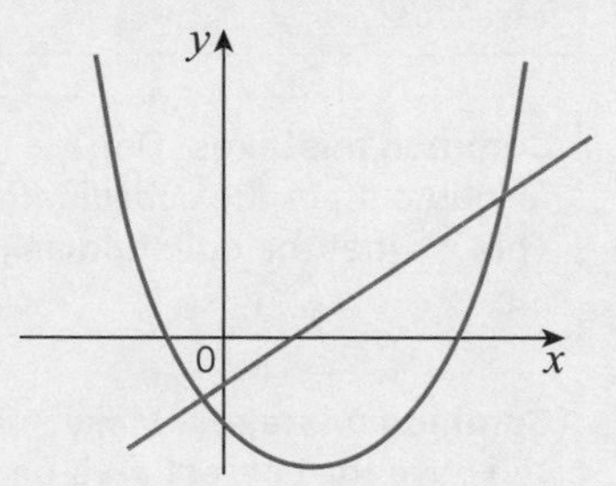

Two points of intersection

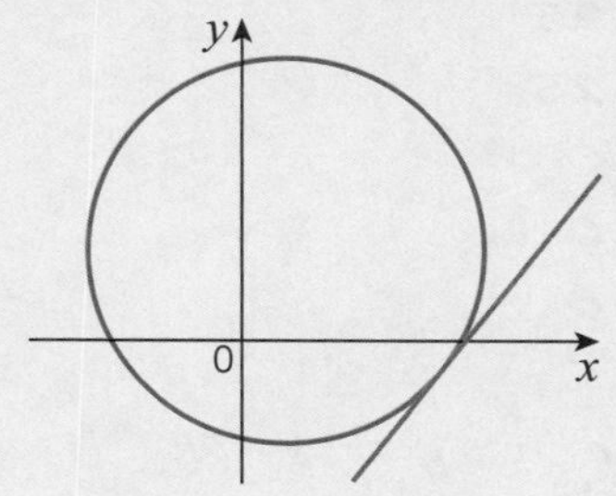

One point of intersection (a tangent)

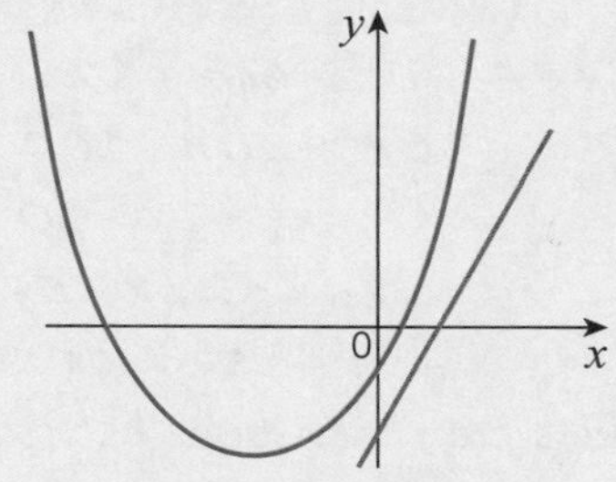

No points of intersection

4 To find the **intersection of two curves** you should substitute the equation for one of the curves into the equation for the other one.

Worked examples

1 Finding the intersection of two lines

Sketch the lines $x + 3y = 11$ and $3x + 2y = 12$ on the same axes, and find the coordinates of the point where they intersect.

Solution

$3x + 2y = 12$

$x + 3y = 11$

(1): $x + 3y = 11$ $\quad 3 \times$ (1): $3x + 9y = 33$

(2): $3x + 2y = 12$ $\quad$ (2): $3x + 2y = 12$

Subtract: $\quad 7y = 21 \Rightarrow y = 3$

Substitute $y = 3$ into (1): $\quad x + 9 = 11 \Rightarrow x = 2$

The coordinates of the point of intersection are (2, 3).

Substituting $x = 0$ into $3x + 2y = 12$ gives $2y = 12 \Rightarrow y = 6$.
So the line goes through (0, 6).

$x + 3y = 11$ goes through (11, 0) and $\left(0, \frac{11}{3}\right)$.

Substituting $y = 0$ into $3x + 2y = 12$ gives $3x = 12 \Rightarrow x = 4$.
So the line goes through (4, 0).

Hint: Check you have the right answer by substituting the values into the other equation:

$$3x + 2y = 3 \times 2 + 2 \times 3$$
$$= 6 + 6$$
$$= 12$$

2 Finding the intersection of a line and a curve

Find the coordinates of the two points where the line $y = 3x - 7$ intersects the circle $(x-2)^2 + (y-4)^2 = 25$.

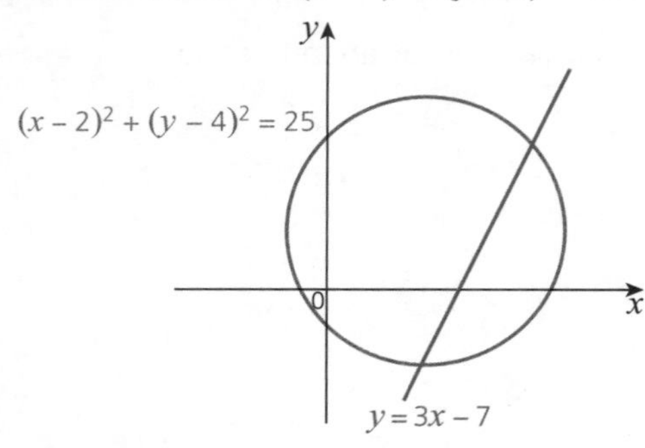

> **Hint:** You should always substitute the linear equation into the equation for the curve.

> **Hint:** You should use the linear equation to find the y-values once you have found the x-values.

> **Common mistakes:** Don't forget to state the coordinates. This is what the question asked for.

> **Common mistakes:** Make sure you have the correct y-value paired-up with the correct x-value.

Solution

Substitute $y = 3x - 7$ into $(x-2)^2 + (y-4)^2 = 25$:

$$(x-2)^2 + ((3x-7)-4)^2 = 25$$
$$\Rightarrow \quad (x-2)^2 + (3x-11)^2 = 25$$
$$\Rightarrow x^2 - 4x + 4 + 9x^2 - 66x + 121 = 25$$
$$\Rightarrow \quad 10x^2 - 70x + 100 = 0$$
$$\Rightarrow \quad x^2 - 7x + 10 = 0$$
$$\Rightarrow \quad (x-2)(x-5) = 0$$
$$\Rightarrow \quad x = 2 \text{ or } x = 5$$

These values can then be substituted into $y = 3x - 7$:

$y = 6 - 7 = -1$ and $y = 15 - 7 = 8$

So the points of intersection are $(2, -1)$ and $(5, 8)$.

3 The tangent to a curve

Show that the line $y = 2x - 7$ is a tangent to the curve $y = x^2 - 2x - 3$ and give the coordinates of the point where the line and curve meet.

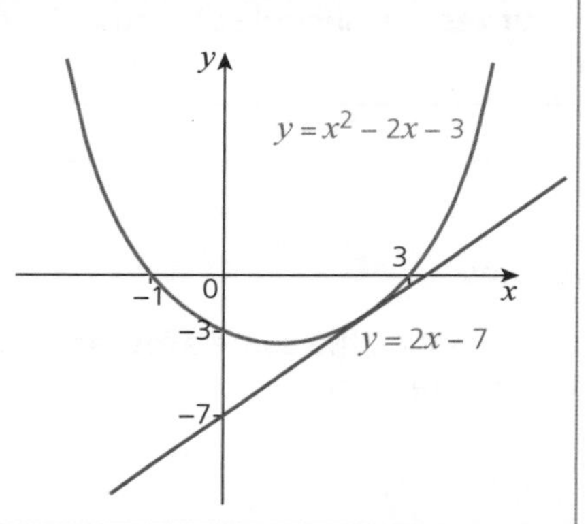

Solution

Substituting $y = 2x - 7$ into $y = x^2 - 2x - 3$

> You should always substitute the linear equation into the equation for the curve.

$2x - 7 = x^2 - 2x - 3$

$\Rightarrow \quad 0 = x^2 - 4x + 4$

> This is a perfect square.

$\Rightarrow \quad 0 = (x-2)(x-2)$

$\Rightarrow \quad x = 2$ (repeated root)

Substituting $x = 2$ into $y = 2x - 7$ gives:

$$y = 4 - 7$$
$$\Rightarrow y = 4 - 3$$

The coordinates of the point of intersection are $(2, -3)$.

There is a repeated (single) point of intersection and therefore the line is a tangent.

4 Solving problems involving circles

The circle $(x+1)^2+(y-2)^2=16$ is shown in the figure.

i Find the equations of the two tangents to the circle which are parallel to the y-axis.

ii Show that the line $y=x-4$ does not intersect the circle.

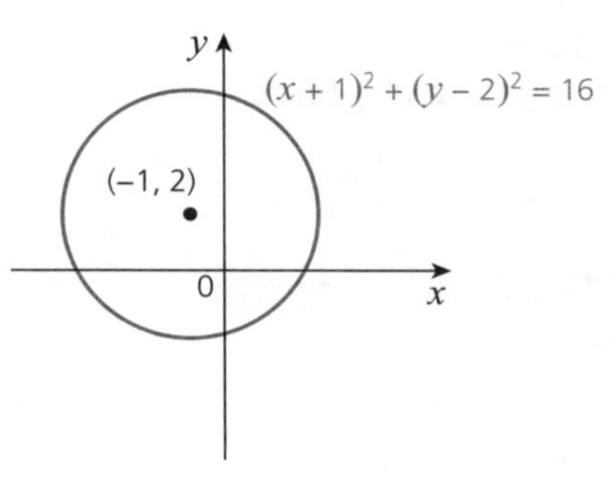

Solution

i The centre of the circle is at $(-1, 2)$.

The radius of the circle is $\sqrt{16}=4$.

The lines are both distance 4 from the centre.

The equations of the tangents are:

$x=-5$ and $x=3$.

The tangents parallel to the y-axis are vertical lines of the form $x=$ **constant**.

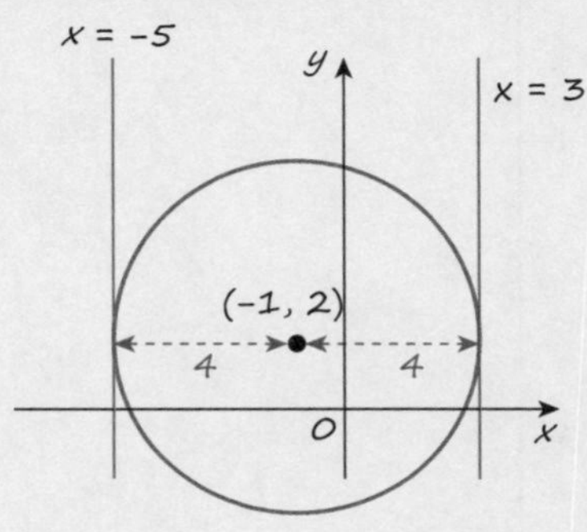

ii Substitute $y=x-4$ into $(x+1)^2+(y-2)^2=16$

$$(x+1)^2+(x-4-2)^2=16$$

$$\Rightarrow \quad (x+1)^2+(x-6)^2=16$$

$$\Rightarrow \quad x^2+2x+1+x^2-12x+36-16=0$$

$$\Rightarrow \quad 2x^2-10x+21=0$$

This is a quadratic equation, so expand it and set it equal to 0.

Find the value of the discriminant

$$b^2-4ac=(-10)^2-4\times2\times21$$
$$=100-168$$
$$=-68$$

The discriminant is the expression under the square-root in the quadratic equation formula:

$$x=\frac{-b\pm\sqrt{b^2-4ac}}{2a}$$

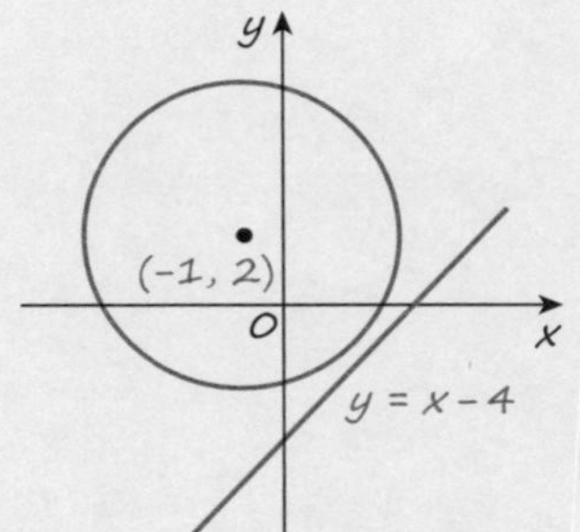

When $b^2-4ac<0$ then $\sqrt{b^2-4ac}$ is not a real number and so there are no real solutions.

Because the discriminant is negative the quadratic equation has no real solutions, so the line and circle do not intersect.

5 Finding the intersection of two curves

Show that the graphs of $y = 2x^2 - 8x + 5$ and $y = x^2 - 4x + 2$ intersect twice and find the coordinates of the two points of intersection.

Solution

Equating the two expressions for y:

This just means that you put the two expressions for y equal to each other.

$$2x^2 - 8x + 5 = x^2 - 4x + 2$$

$$\Rightarrow \quad x^2 - 4x + 3 = 0$$

$$\Rightarrow \quad (x - 1)(x - 3) = 0$$

$$\Rightarrow \quad x = 1 \text{ or } x = 3$$

Substituting the x coordinates into $y = x^2 - 4x + 2$:

$x = 1 \Rightarrow y = 1 - 4 + 2 = -1$

$x = 3 \Rightarrow y = 9 - 12 + 2 = -1$

So there are two points of intersection at $(1, -1)$ and $(3, -1)$.

You can check your answer is correct by substituting the x-coordinates into $y = 2x^2 - 8x + 5$.

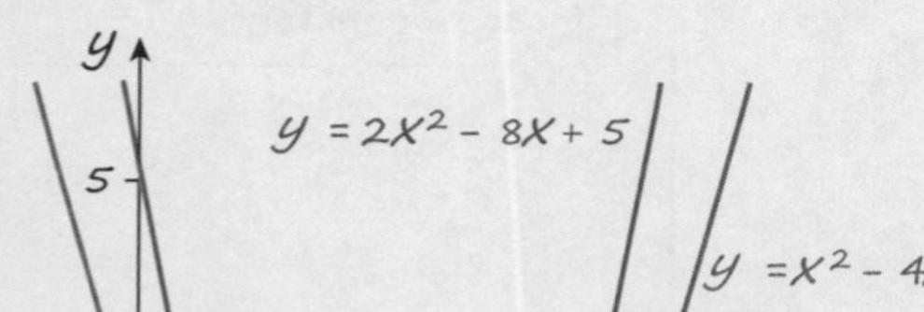

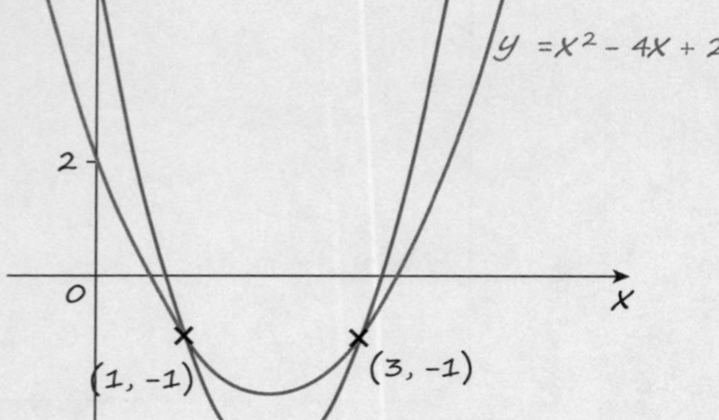

Test yourself

TESTED

1 Find the coordinates of the point where the lines $2x + 3y = 12$ and $3x - y = 7$ intersect.

A $\left(\frac{9}{7}, \frac{22}{7}\right)$ B (3, 2) C (9, 2) D (3, 6) E $\left(\frac{33}{8}, \frac{5}{4}\right)$

2 Find the coordinates of the points where the line $y = 2x + 3$ intersects the curve $y = x^2 + 3x + 1$.

A (–2, 5) and (1, –1) B (–4, –5) and (–1, 1) C (–2, –9) and (1, 5)

D (–1, 1) and (2, 7) E (–2, –1) and (1, 5)

3 Find the equations of the two tangents to the circle $(x - 2)^2 + (y + 1)^2 = 9$ which are parallel to the x-axis.

A $x = -5$ and $x = 1$ B $x = 5$ and $x = -1$ C $y = 8$ and $y = -10$

D $y = 4$ and $y = -2$ E $y = 2$ and $y = -4$

4 Which of the following lines does **not** intersect the circle $x^2 - 8x + y^2 - 9 = 0$ shown in the diagram.

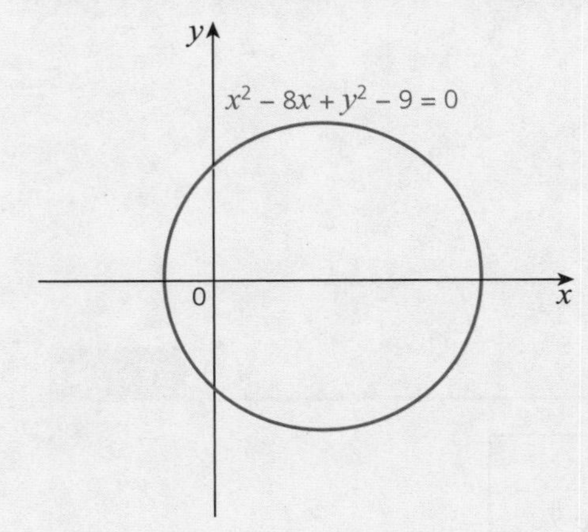

A $y = 2x + 2$

B $y = -5$

C $y = 2x + 4$

D $y = x - 1$

E $x = 7$

5 How many points of intersection are there between the curve $y = x^2 - 3$ and the circle $x^2 + y^2 = 9$?

A 0 B 1 C 2 D 3 E 4

Full worked solutions online

CHECKED ANSWERS

Exam-style question

i Find the coordinates of the points where the graph of $y = x^2 + 2x - 3$ crosses the axes and sketch the graph.

ii Show that the graphs of $y = x^2 + 2x - 3$ and $y = x^2 - 10x + 21$ intersect only once and find the coordinates of the point of intersection.

iii Show that the line $y = -2x - 8$ does not intersect either curve.

Short answers on page 154

Full worked solutions online

CHECKED ANSWERS

Chapter 6 Trigonometry

About this topic

You will often need to solve equations involving trigonometrical functions. By sketching their curves you will be able to find an angle of any size.

Another common task you will have is to find an angle or a length of side of a triangle without right angles. The sine and cosine rules allow you to do this.

Before you start, remember

- the trigonometrical ratios: sine, cosine and tangent
- surds – see Chapter 2
- bearings.

Working with trigonometric functions

REVISED

Key facts

1 Trigonometrical functions for values of angle θ between 0° and 90° inclusive.

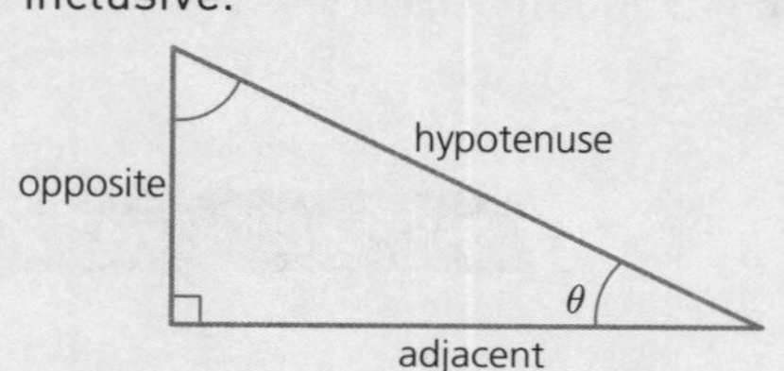

$$\sin\theta = \frac{\text{opposite}}{\text{hypotenuse}} = \frac{O}{H}$$

$$\cos\theta = \frac{\text{adjacent}}{\text{hypotenuse}} = \frac{A}{H}$$

$$\tan\theta = \frac{\text{opposite}}{\text{adjacent}} = \frac{O}{A}$$

2 Sometimes you will be asked to give the answers in surd form. So you need to know the trigonometrical ratios of special angles:

i equilateral triangle

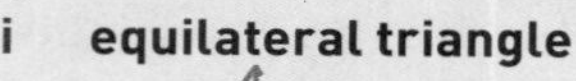

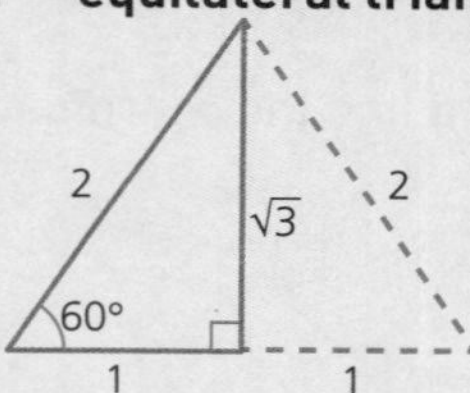

$$\sin 60° = \frac{O}{H} = \frac{\sqrt{3}}{2}$$

$$\cos 60° = \frac{A}{H} = \frac{1}{2}$$

$$\tan 60° = \frac{O}{A} = \frac{\sqrt{3}}{1} = \sqrt{3}$$

ii equilateral triangle

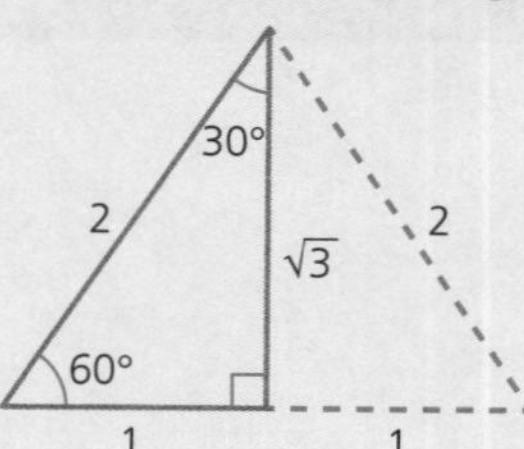

$$\sin 30° = \frac{O}{H} = \frac{1}{2}$$

$$\cos 30° = \frac{A}{H} = \frac{\sqrt{3}}{2}$$

$$\tan 30° = \frac{O}{A} = \frac{1}{\sqrt{3}} = \frac{\sqrt{3}}{3}$$

iii isosceles triangle

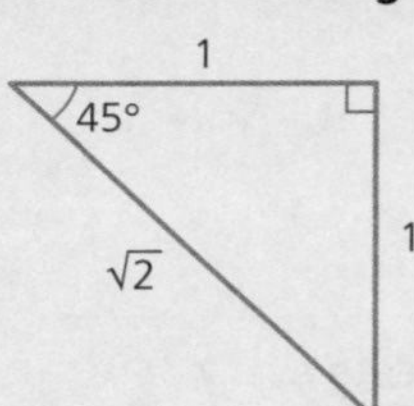

$$\sin 45° = \frac{O}{H} = \frac{1}{\sqrt{2}} = \frac{\sqrt{2}}{2}$$

$$\cos 45° = \frac{A}{H} = \frac{1}{\sqrt{2}} = \frac{\sqrt{2}}{2}$$

$$\tan 45° = \frac{O}{A} = \frac{1}{1} = 1$$

3 The unit circle is a circle with radius 1 unit. The following are true for any point $P(x, y)$ on the unit circle and acute angle θ between OP and the x-axis:

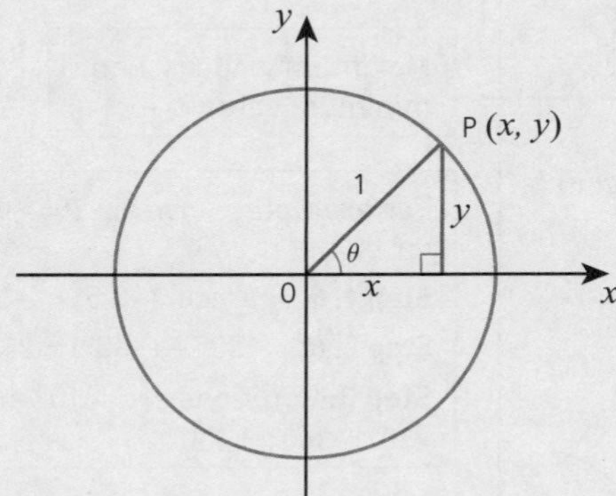

$$\sin\theta = \frac{y}{1} = y$$

$$\cos\theta = \frac{x}{1} = x$$

$$\tan\theta = \frac{y}{x}$$

$$y^2 + x^2 = 1$$

Using Pythagoras' theorem.

$$\sin 0° = 0$$

$$\cos 0° = 1$$

Two important identities are

$$\sin^2\theta + \cos^2\theta \equiv 1$$

and $$\tan\theta \equiv \frac{\sin\theta}{\cos\theta}, \quad \cos\theta \neq 0$$

4 The angles in the anticlockwise direction from the x-axis are positive and in the clockwise direction are negative.

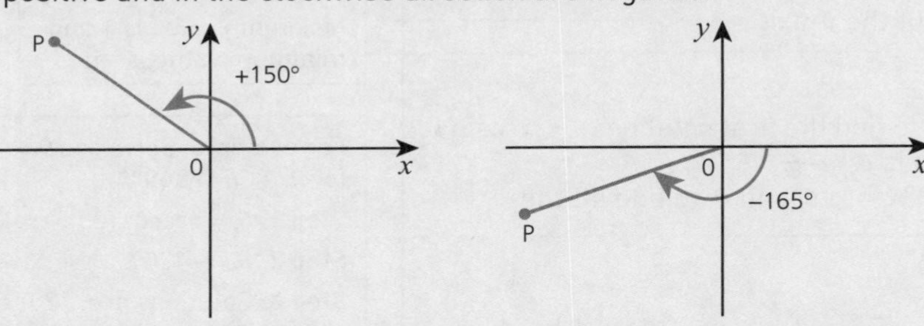

5 You can use a CAST diagram to decide whether the sin, cos or tan of your angle is positive or negative.

2nd quadrant: $x < 0$ and $y > 0$
Only sin θ is positive.

1st quadrant: $x > 0$ and $y > 0$
sin θ, cos θ and tan θ are **all positive**.

y, +, S, A, −, +, x, T, C, −

3rd quadrant: $x < 0$ and $y < 0$
Only tan θ is positive.

4th quadrant: $x > 0$ and $y < 0$
Only cos θ is positive.

6 You can use the properties of the graphs of $y = \sin\theta$, $y = \cos\theta$ and $y = \tan\theta$ to help you solve equations.

7 Graph of $y = \sin\theta$

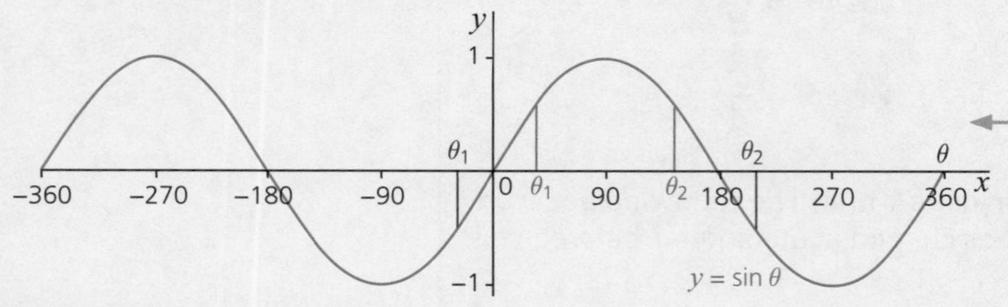

It repeats every 360°.

- Period of $y = \sin\theta$ is 360°
- $y = \sin\theta$ has rotational symmetry of order 2 about the origin
- $-1 \leqslant \sin\theta \leqslant 1$
- To solve $\sin\theta = k$:

  Step 1: Use your calculator to find the first solution: $\theta_1 = \arcsin k$.

  Step 2: The second solution is $\theta_2 = 180° - \theta_1$.

  Step 3: Add or subtract 360° from θ_1 and θ_2 to find all the solutions.

$\sin\theta = -\sin(-\theta)$
e.g. $\sin 30° = -\sin(-30°)$.

Maximum value is 1 and minimum value is –1.

For example: Solve $\sin\theta = -0.5$ for $0° \leqslant \theta \leqslant 360°$.

Step 1: $\theta_1 = \arcsin(-0.5) = -30°$.

Step 2: $\theta_2 = 180° - (-30°) = 210°$.

Step 3: Solutions are: 210° and $-30° + 360° = 330°$.

8 Graph of $y = \cos\theta$

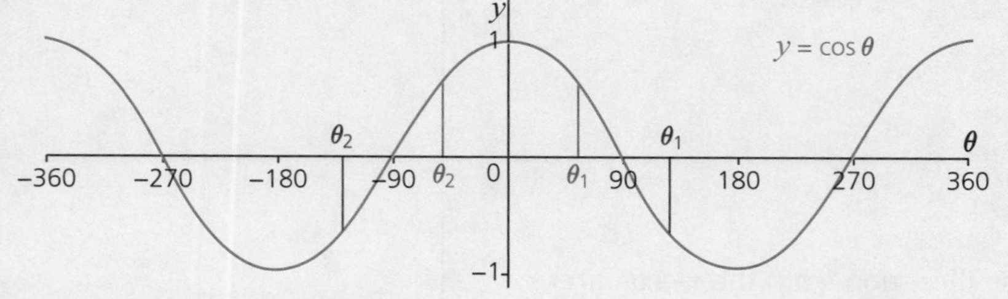

- Period of $y = \cos\theta$ is 360°
- $y = \cos\theta$ is symmetrical about the y-axis
- $-1 \leqslant \cos\theta \leqslant 1$
- To solve $\cos\theta = k$:

  Step 1: Use your calculator to find the first solution: $\theta_1 = \arcsin k$.

  Step 2: The second solution is $\theta_2 = -\theta_1$.

  Step 3: Add or subtract 360° from θ_1 and θ_2 to find all the solutions.

$\cos\theta = \cos(-\theta)$
e.g. $\cos 30° = \cos(-30°)$.

Maximum value is 1 and minimum value is –1.

For example: Solve $\cos\theta = -0.5$ for $0° \leqslant \theta \leqslant 360°$.

Step 1: $\theta_1 = \arccos(-0.5) = 120°$.

Step 2: $\theta_2 = -120°$.

Step 3: Solutions are: 120° and $-120° + 360° = 240°$.

9 Graph of $y = \tan\theta$

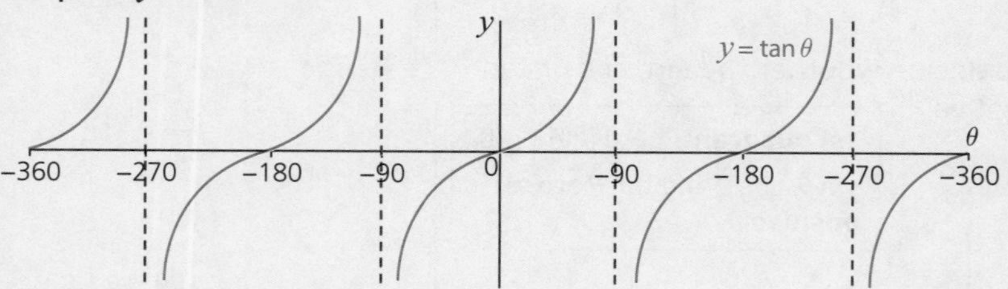

- Period of $y = \tan\theta$ is 180°
- $y = \tan\theta$ has rotational symmetry of order 2 about the origin
- There are asymptotes at $x = \pm 90°$, $x = \pm 270°$...
- To solve $\tan\theta = k$:

  Step 1: Use your calculator to find the first solution: $\theta_1 = \arctan k$.

  Step 2: Add or subtract 180° from θ_1 to find all the solutions.

For example: Solve $\tan\theta = -0.5$ for $\leqslant \theta \leqslant 360°$.

Step 1: $\theta_1 = \arctan(-0.5) = -26.6°$.

Step 2: Solutions are: $-26.6° + 180° = 153.4°$ and $153.4° + 180° = 333.4°$.

Worked examples

1 Using exact values

Without using a calculator work out $\sin^2 30° - \cos^2 30° \tan^2 30°$.

$\sin^2 30°$ means $(\sin 30°)^2$.

Solution

Using the trigonometrical ratios for special triangles

$$\sin 30° = \frac{1}{2}, \quad \cos 30° = \frac{\sqrt{3}}{2} \quad \text{and} \quad \tan 30° = \frac{1}{\sqrt{3}}$$

$$\text{So } \sin^2 30° - \cos^2 30° \tan^2 30° = \left(\frac{1}{2}\right)^2 - \left(\frac{\sqrt{3}}{2}\right)^2 \times \left(\frac{1}{\sqrt{3}}\right)^2$$

$$= \frac{1}{4} - \frac{3}{4} \times \frac{1}{3} = 0.$$

Use brackets to show that both the top line and the bottom line of the fraction are squared.

2 Using the CAST diagram

Given that $\cos\theta = \frac{12}{13}$ and θ is reflex, find the exact value of:

i $\sin\theta$ ii $\tan\theta$.

Solution

i *Using the identity* $\sin^2\theta + \cos^2\theta \equiv 1$

$$\Rightarrow \sin^2\theta \equiv 1 - \cos^2\theta$$

$$\Rightarrow \sin^2\theta = 1 - \left(\frac{12}{13}\right)^2 = 1 - \frac{144}{169} = \frac{25}{169}$$

$$\Rightarrow \sin\theta = \pm\frac{5}{13}$$

You need to work out if $\sin\theta$ is positive or negative.

Using the CAST diagram, when $\cos\theta$ is positive and θ is reflex then θ is in the fourth quadrant and $\sin\theta$ and $\tan\theta$ are negative.

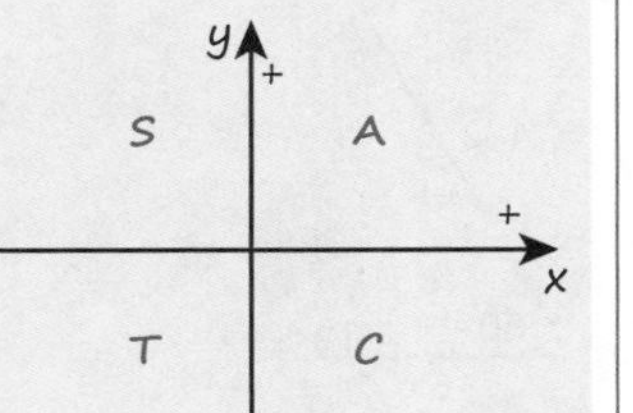

So $\sin\theta = -\frac{5}{13}$.

ii *Using the identity* $\tan\theta \equiv \frac{\sin\theta}{\cos\theta}$

$$\Rightarrow \tan\theta = \frac{-\frac{5}{13}}{\frac{12}{13}} = -\frac{5}{12}$$

To divide by a fraction, you multiply by its reciprocal: $-\frac{5}{13} \times \frac{13}{12} = -\frac{5}{12}$.

3 Proving identities

Prove that $\frac{\cos^2\theta - 1}{\sin^2\theta - 1} \equiv \tan^2\theta$.

Solution

$$\frac{\cos^2\theta - 1}{\sin^2\theta - 1} \equiv \frac{\sin^2\theta}{\cos^2\theta}$$

$\equiv \tan^2\theta$ *as required.*

Using the identity $\sin^2\theta + \cos^2\theta \equiv 1$.

4 Solving trigonometric equations (1)

Solve the equation for $\cos^2\theta - 2\cos\theta = 0$ for $0° \leqslant \theta \leqslant 360°$.

Solution

$\cos^2\theta - 2\cos\theta = 0$

$\Rightarrow \cos\theta(\cos\theta - 2) = 0$

Either $\cos\theta = 0$ or $\cos\theta = 2$ this is impossible.

$\Rightarrow \theta = 90°$ or $\theta = 270°$

Common mistake: Factorise the left-hand side. Don't divide by $\cos\theta$, otherwise you will lose the solutions to $\cos\theta = 0$.

The maximum value of $\cos\theta$ is 1 and the minimum value is –1.

$-90° + 360° = 270°$.

5 Solving trigonometric equations (2)

Solve $\sin(x - 65°) = \frac{1}{2}$ for $0° \leqslant x \leqslant 360°$.

Solution

Let $\theta = x - 65°$

$\theta = \arcsin\left(\frac{1}{2}\right) = 30°$ or $\theta = 150°$

So $x - 65° = 30°$ or $x - 65° = 150°$

$x = 95°$ $x = 215°$

Test yourself

TESTED

1 Four of these diagrams below are correct. Which one of the diagrams is incorrect?

A

$\sqrt{7}$, 60°, $2\sqrt{7}$, $\sqrt{21}$

B

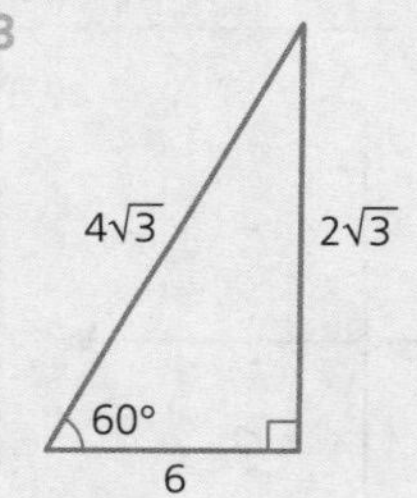

C

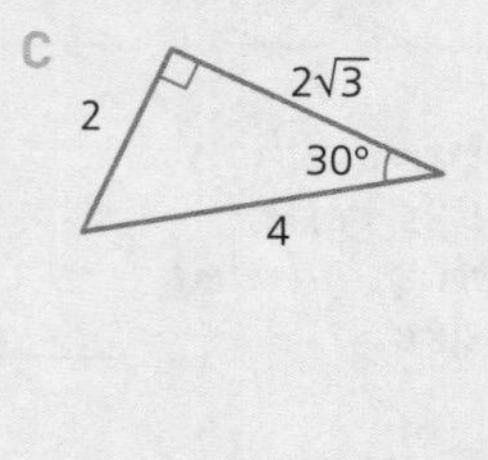

D

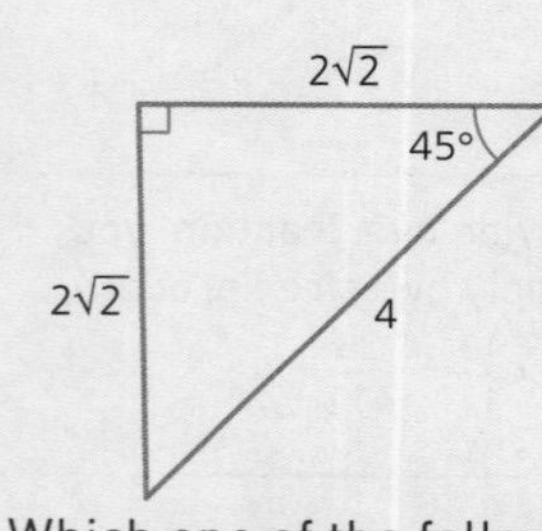

E

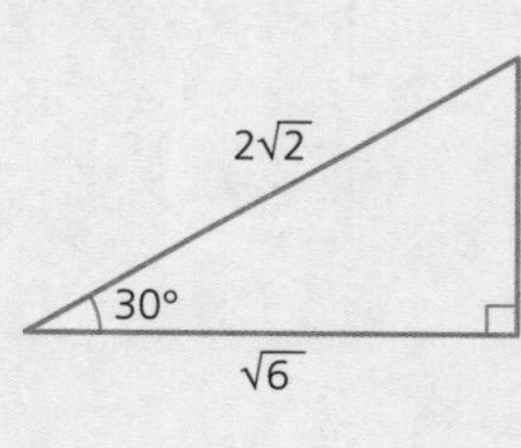

2 Which one of the following has the same value as $\tan 315°$?

A $\tan(-315°)$ B $-\tan 45°$ C $\tan 225°$ D $\tan 45°$

3 Solve the equation $\cos 2x = -0.5$ to the nearest degree for $0° \leqslant x \leqslant 360°$.

A 60° B 60°, 120°, 240° and 300° C 120° and 240° D 120° E 60°, 120° and 240°

4 Solve the equation $\sin^2\theta = \sin\theta$ to the nearest degree for $0° \leqslant \theta \leqslant 360°$.

A 0° and 90° B 0°, 90°, 180° and 360° C 0°, 180° and 360° D 90°

5 Which one of the following is the exact value of $\dfrac{1 - \sin 240°}{1 + \sin 240°}$?

A $7 - 4\sqrt{3}$ B 1 C $7 + 4\sqrt{3}$ D 3 E 7

Full worked solutions online

CHECKED ANSWERS

Exam-style question

i Express $2\cos^2 x - 3\sin x$ as a quadratic function of $\sin x$.
ii Hence solve the equation $2\cos^2 x = 3\sin x$ for $0° \leqslant x \leqslant 360°$.

Short answers on page 154

Full worked solutions online

CHECKED ANSWERS

Triangles without right angles

REVISED

Key facts

1 Usually the vertices of any triangle are labelled with capital letters, and the opposite sides with corresponding small letters.

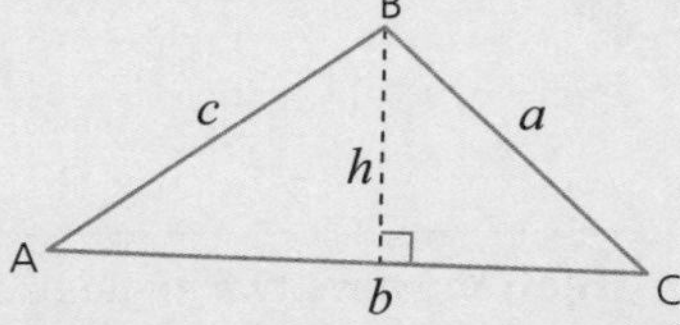

2 The area of triangle ABC is $\frac{1}{2}ab\sin C$

You can find the area of any triangle if you know two sides and the angle between them.

3 The sine rule for triangle ABC is

$$\frac{\sin A}{a} = \frac{\sin B}{b} = \frac{\sin C}{c}$$

or $$\frac{a}{\sin A} = \frac{b}{\sin B} = \frac{c}{\sin C}$$

Use this form to find a missing angle ...

... and this form to find a missing side.

When you use the sine rule to find a missing angle, θ, always check whether $180° - \theta$ is also a solution.

This is called the **ambiguous case**.

4 The cosine rule for the triangle ABC is

$$a^2 = b^2 + c^2 - 2bc\cos A$$

or $$\cos A = \frac{b^2 + c^2 - a^2}{2bc}$$

Use the cosine rule when you know:

- Two sides and the angle between them and you need the third side.
- All three sides and you need to find any angle.

Worked examples

1 Find the area of a triangle

Find the area of the triangle ABC shown below:

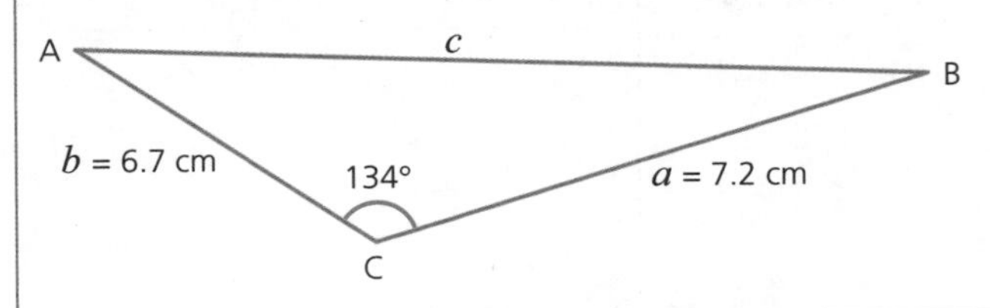

Solution

$\text{Area} = \frac{1}{2}ab\sin C$

$\text{Area} = \frac{1}{2} \times 7.2 \times 6.7 \times \sin 134°$

$= 17.35...$

$\text{Area} = 17.4\,\text{cm}^2$ (to 3 s.f.)

Be careful with the units.

2 Using the sine rule to find a missing side

Find the side c in the triangle below. Give your answer to the nearest 0.1 cm.

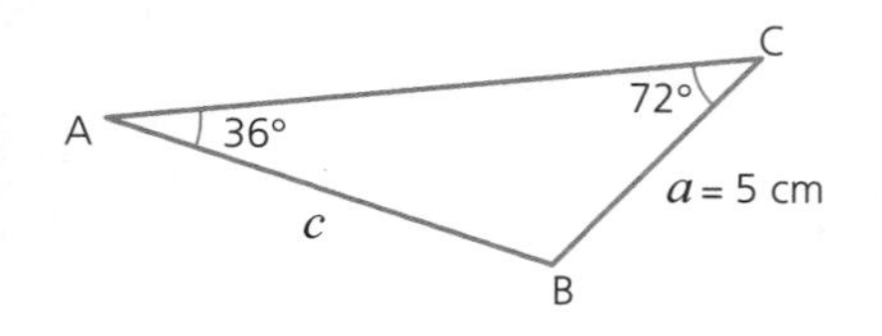

Common mistake: Make sure that your calculator is set in degrees mode.

Solution

Hint: You have two angles and one side so use the sine rule.

You are given two angles: $\angle A = 36°$ and $\angle C = 72°$.

You are also given that $BC = a = 5\,\text{cm}$.

Use the sine rule in the form: $\frac{c}{\sin C} = \frac{a}{\sin A}$

$\frac{c}{\sin 72°} = \frac{5}{\sin 36°}$

$c = \sin 72° \times \frac{5}{\sin 36°}$

$= 8.090...$

Using your calculator.

$c = 8.1$ cm (nearest 0.1 cm)

Round off only in the final answer.

3 Using the sine rule to find a missing angle

In triangle XZY, XZ = 3.6 cm, YZ = 4.5 cm and $\angle Y = 47°$.

i Draw the triangle.

ii Find the possible size of angle X.

Solution

i When you draw the triangle you find there are two possible angles and two possible positions for point X. These are marked X_1 and X_2.

Hint: You have two sides and a non-included angle, so use the sine rule.

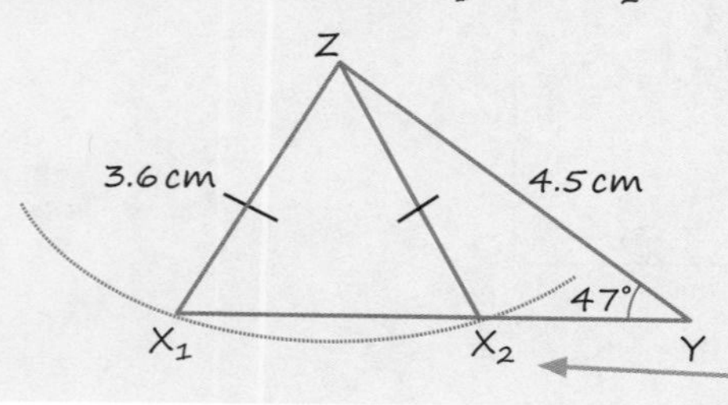

You can see from the diagram that X lies on the circle with centre Z and radius of 3.6 cm.

ii $\frac{\sin X}{4.5} = \frac{\sin 47°}{3.6}$

$\sin X = 4.5 \times \frac{\sin 47°}{3.6}$

$\sin X = 0.91419...$

$X = 66.091...°$ or $180° - 66.091...°$

So $\angle X_1 = 66.1°$ or $\angle X_2 = 180° - 66.1° = 113.9°$

(both to the nearest 0.1°)

Hint: Always check for the ambiguous case. There are two possible values of X between 0° and 180°.

4 Using the cosine rule to find a missing angle

Find the angle ABC in the triangle shown below:

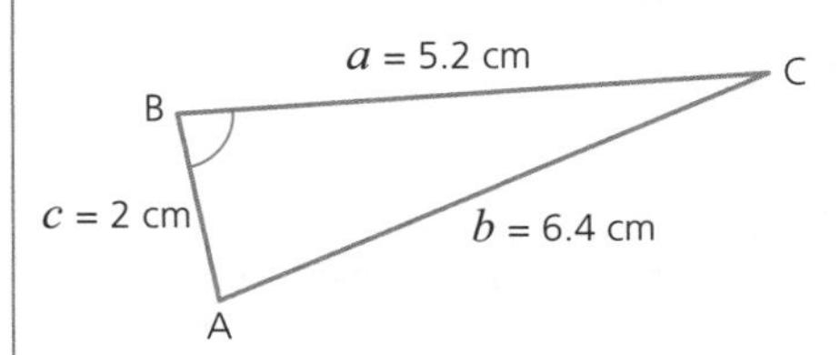

Solution

Using cosine rule:

$\cos B = \frac{c^2 + a^2 - b^2}{2ca}$

$\cos B = \frac{2^2 + 5.2^2 - 6.4^2}{2 \times 2 \times 5.2} = -0.4769...$

$B = 118°$ (to the nearest degree)

In this example, three sides are given and you need to find a missing angle so you should use the cosine rule.

5 Using the cosine rule to find a missing side

Find the length of the side p in the triangle below.

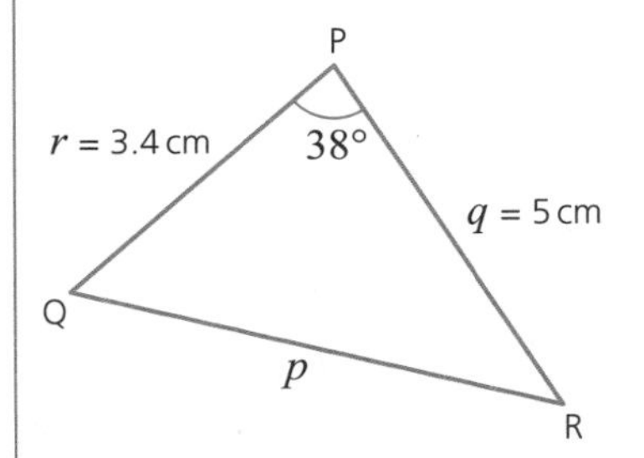

Solution

In this case use the cosine rule in the form:

$p^2 = q^2 + r^2 - 2qr \cos P$

$p^2 = 5^2 + 3.4^2 - 2 \times 5 \times 3.4 \times \cos 38°$

$\Rightarrow p^2 = 9.7676...$

$\Rightarrow p = 3.125... = 3.13$ cm (to 3 s.f.)

Common mistake: Don't round until you get to your final answer.

Keep all other values stored in your calculator.

6 Solving problems involving triangles without right angles

Ship A is 3 km from a lighthouse L on a bearing of 137°.

Ship B is 6.5 km from the lighthouse on a bearing of 067°.

Find the distance and bearing of the ship A from the ship B.

Solution

You need to find the distance AB and the angle NBA.

Start by drawing a diagram and filling in all of the information you can.

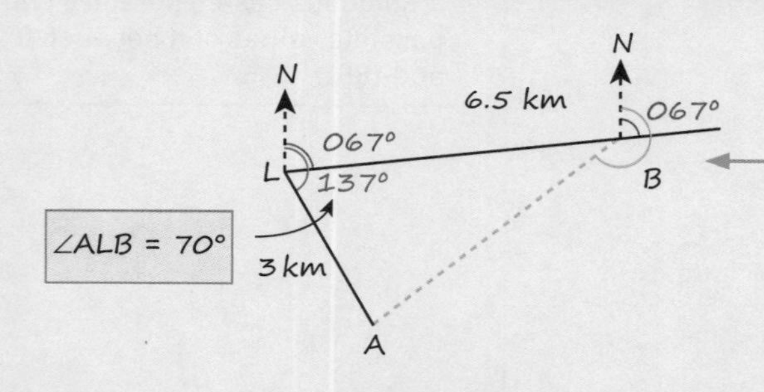

Notice that $\angle NBA = 360^\circ - (\angle ABL + \angle LBN)$. So start by finding $\angle ABL$ and $\angle LBN$.

First look at angle ALB. It is 137° − 67° = 70°

Now use the cosine rule in the triangle ALB.

You know two sides and the included angle, so use the cosine rule to find the third side.

$AB^2 = 6.5^2 + 3^2 - 2 \times 6.5 \times 3 \cos 70^\circ$

$AB = 6.1572...$

Don't round yet.

The bearing of ship A from ship B is angle NBA (shown in green).

First find ∠ABL using sine rule $\frac{\sin B}{3} = \frac{\sin 70^\circ}{6.1572...}$

You can use three dots to show that you have used the unrounded value in your working.

$\sin B = 3 \times \frac{\sin 70^\circ}{6.1572...}$

$= 0.4578...$

$\angle ABL = 27.248...$

The angle LBN is 180° − 67° = 113°

This is 360° − (113° + 27.25°) = 219.75°.

The bearing is 220° to the nearest degree.

Test yourself

TESTED

1 In the triangle ABC, ∠CAB = 37°, ∠ABC = 56° and CB = 4 cm. Find the length of AC.

A 3.24 cm B 8.02 cm C 5.51 cm D 0.18 cm E 2.90 cm

2 In the triangle XYZ, XY = 3.8 cm, YZ = 4.5 cm and ∠YZX = 40°. Three of the following statements are false and one is true. Which one is true?

A A possible value for the area of the triangle is exactly 8.55 cm².

B The only possible value of ∠XYZ is 90° to the nearest degree.

C You can find the remaining side and angles of the triangle using only the cosine rule.

D The possible values of ∠YXZ are 50° and 130° (to the nearest degree).

3 In the triangle MNP, MN = 5.4 cm, NP = 6 cm and MP = 7 cm. Find angle MNP correct to 3 s.f.

A 48.3° B 75.6° C 56.1° D 1.32°

4 For the triangle given below three of the statements are true and one is false. Which one is false?

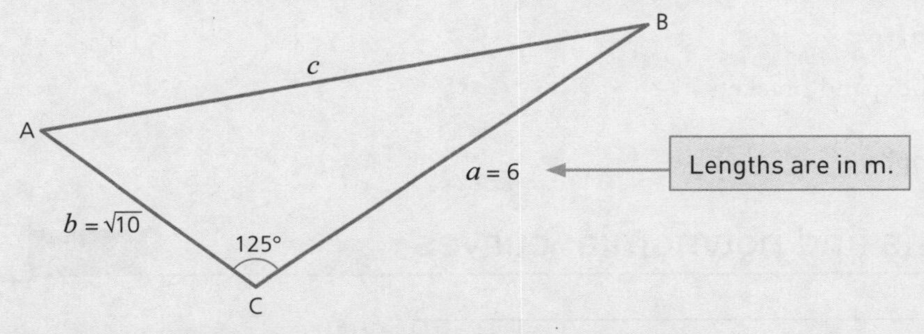

A The area of the triangle is 7.77 m² (to 3 s.f.).

B AB is 8.23 m (to 3 s.f.).

C Using only the sine rule you can find the value of c.

D ∠B = 18.34° (to 2 d.p.).

5 At 12 noon a ship is at point M which is on a bearing of 148° from a lighthouse, L. The ship travels due East at 20 km per hour and at 1230 hours it is at point N, on a bearing of 127° from the lighthouse.

Three of the following statements are false and one is true. Which one is true?

A At 1230 hours the ship is 17 km from the lighthouse.

B The area LMN is 100 km², to the nearest whole number.

C At noon the ship is 16.8 km from L to the nearest km.

D If the ship continues on the same speed and the same course, it will be on a bearing of 106° from the lighthouse at 1300 hours.

Full worked solutions online

CHECKED ANSWERS

Exam-style question

In a quadrilateral ABCD, AD = 7 cm, DC = 5 cm, ∠ADC = 47°, ∠ABC = 127° and ∠BAC = 35°.

i Find the length of AC.

ii Find the angle CAD.

iii Find the length of AB.

iv Find the area of the quadrilateral.

Short answers on page 154

Full worked solutions online

CHECKED ANSWERS

Chapter 7 Polynomials

About this topic

'Poly' is a Greek word and means many. Polynomials are expressions involving many terms with positive integer powers; you need to be able to add, subtract, multiply and divide polynomials.

You also need to use the factor theorem to factorise a polynomial and sketch polynomial curves.

Before you start, remember

- how to simplify expressions and expand brackets
- how to factorise expressions
- quadratic functions and their graphs.

Polynomial expressions and polynomial curves

REVISED

Key facts

1. The **order** of the polynomial is the highest power of the variable it contains. So $4 - 7x^5 + 3x^{12}$ has order 12.
2. When you **multiply** polynomials remember:
   - x means x^1
   - when you multiply powers of x add the indices: $x^3 \times x^5 = x^{3+5} = x$
   - the rules for multiplying positive and negative numbers are:
     $+ \times + = +$; $- \times - = +$; $+ \times - = -$; $- \times + = -$
3. Polynomial curves have **turning points**.

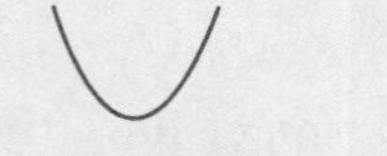
Minimum point

Maximum point

4. The **number of turning points** depends on the order of a polynomial.

Order of a polynomial	Maximum number of turning points	Notes
1	0	Straight line
2	1	Quadratic
3	2	Cubic
4	3	Quartic
...	...	...
n	$(n - 1)$	

There may be fewer turning points, depending on the equation of the curve.

5 To **sketch** the curve of a polynomial you have to:

i Decide on the shape of the curve by looking at the highest power of x, ax^n.

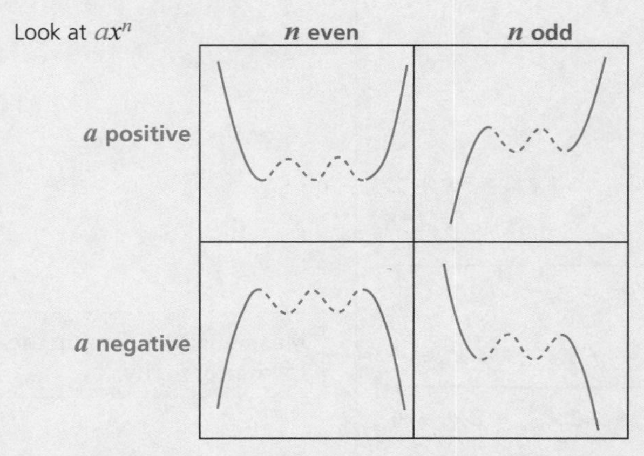

ii Show the turning points. Sometimes you will be asked to write the coordinates of the turning points.

iii Give the coordinates of the points where the curve crosses (intersects) the x-axis and the y-axis.

6 To **plot** the curve of a polynomial you have to be more accurate. Calculate the values of y for suitable values of x, plot these points and join them as a smooth curve.

Hint: Remember for a polynomial of degree n there are at most $(n-1)$ turning points.

Worked examples

1 Adding and subtracting polynomials

You are given that $f(x) = 2x^3 - 7x^2 + 4x - 5$ and $g(x) = x^3 + 9x - 7$.

Find:

i $f(x) + g(x)$

ii $f(x) - g(x)$.

Solution

i *Collecting like terms*

$(2x^3 - 7x^2 + 4x - 5) + (x^3 + 9x - 7)$

$= 2x^3 + x^3 - 7x^2 + 4x + 9x - 5 - 7$

$= 3x^3 - 7x^2 + 13x - 12$

Using columns

$$\begin{array}{rrrrr} & 2x^3 & -7x^2 & +4x & -5 \\ + & x^3 & & +9x & -7 \\ \hline & 3x^3 & -7x^2 & +13x & -12 \end{array}$$

So $f(x) + g(x) = 3x^3 - 7x^2 + 13x - 12$

Remember x^3 means $1x^3$ so the coefficient of x^3 is 1.

Take care: there is no term of x^2 in the second polynomial, so make sure you leave a space.

ii *Collecting like terms*

$(2x^3 - 7x^2 + 4x - 5) - (x^3 + 9x - 7)$

$= 2x^3 - 7x^2 + 4x - 5 - x^3 - 9x + 7$

$= 2x^3 - x^3 - 7x^2 - 4x - 9x - 5 + 7$

$= x^3 - 7x^2 - 5x + 2$

Using columns

$$\begin{array}{rrrrr} & 2x^3 & -7x^2 & +4x & -5 \\ - & x^3 & & +9x & -7 \\ \hline & x^3 & -7x^2 & -5x & +2 \end{array}$$

So $f(x) - g(x) = x^3 - 7x^2 - 5x + 2$

Common mistakes: A '–' sign in front of the brackets means that you have to multiply each term of the brackets by –1 or change the sign in front of each term in the brackets.

Remember $-(-7) = +7$.

Change the signs first.

2 Multiplying polynomials

You are given that $f(x) = 4x^2 - 3x + 2$ and $g(x) = x - 2$.

Find $f(x) \times g(x)$.

Solution

Using columns

$$\begin{array}{r} 4x^2 - 3x + 2 \\ \times \qquad\quad x - 2 \\ \hline \end{array}$$

Multiply the top line by x. $\quad 4x^3 - 3x^2 + 2x$

Multiply the top line by -2. $\quad -8x^2 + 6x - 4$

Add $\quad 4x^3 - 11x^2 + 8x - 4$

Make sure you line up the terms correctly.

Collecting like terms

$$(x-2)\times(4x^2-3x+2) = x\times(4x^2-3x+2) - 2\times(4x^2-3x+2)$$
$$= 4x^3 - 3x^2 + 2x - 8x^2 + 6x - 4$$
$$= 4x^3 + (-3x^2 - 8x^2) + (2x + 6x) - 4$$
$$= 4x^3 - 11x^2 + 8x - 4$$

Common mistakes: Check the signs carefully.

3 Sketching curves (1)

Sketch the curve $y = x^3 - x = x(x-1)(x+1)$.

Solution

The polynomial is of order 3 (odd number) and the coefficient of x^3 is 1, which is a positive number. So you expect the shape of the curve to be:

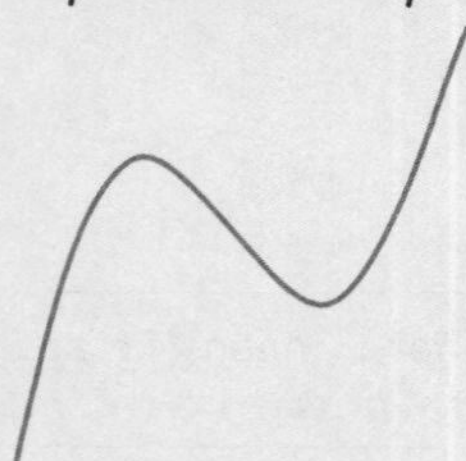

Now find the points where the curve crosses the x-axis and the y-axis:

The curve crosses the x-axis when $y = 0$.

Substitute $y = 0$ into $y = x^3 - x = x(x-1)(x+1)$

and solve the equation $x(x-1)(x+1) = 0$

So $x = 0$ or $x = 1$ or $x = -1$

The curve crosses the x-axis at the points: $(-1, 0)$, $(0, 0)$ and $(1, 0)$

Note that the y coordinates are 0.

The curve crosses the y-axis when $x = 0$.

Substitute $x = 0$ into $y = x^3 - x = 0 - 0 = 0$.

So the curve crosses the y-axis at (0, 0).

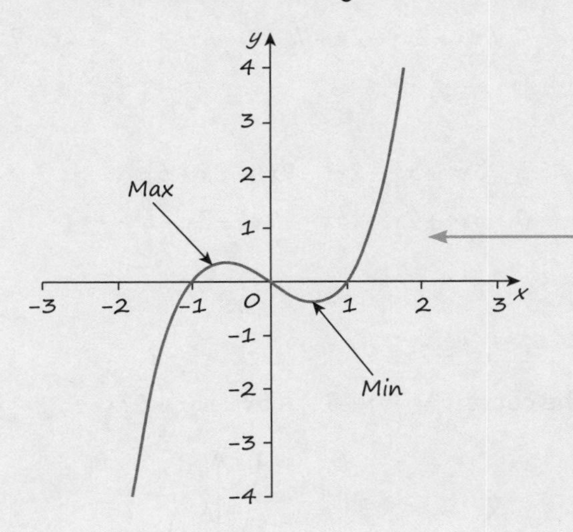

There are two turning points – a maximum and a minimum.

Hint: Make sure you label the points where the curve intersects the axes.

Draw a **smooth** curve.

Use a ruler to draw the axes.

Do not use a ruler to draw the curve.

4 Sketching curves (2)

Sketch the curve $y = -(x - 1)(x - 3)^2$.

Solution

$y = -(x - 1)(x - 3)(x - 3)$

When $y = 0$ $\quad -(x - 1)(x - 3)(x - 3) = 0$

$x = 1$ or $x = 3$ or $x = 3$

When $x = 0$ $\quad y = -(0 - 1)(0 - 3)(0 - 3)$

so $y = 9$

So the curve crosses x-axis at (1, 0).

3 is a repeated root, so the curve just touches the x-axis at (3, 0).

The curve crosses the y-axis at (0, 9).

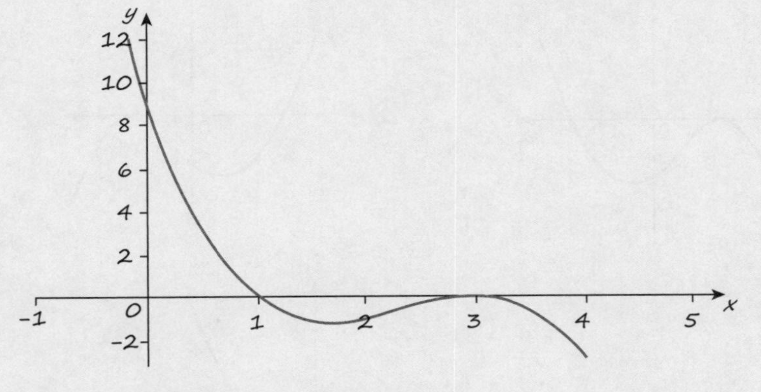

There are two turning points.

Hint: The coefficient of x^3 is negative, so the curve is 'upside-down'.

Test yourself

TESTED

1 You are given that $f(x) = 4x^3 - x + 3$ and $g(x) = 2x^2 - 3x + 4$.

Which of the following polynomials is $f(x) + g(x)$?

A $6x^3 - 4x + 7$ B $4x^3 + 2x^2 - 2x + 7$ C $6x^2 - 4x + 7$ D $4x^3 + 2x^2 + 2x + 7$ E $4x^3 + 2x^2 - 4x + 7$

2 You are given $f(x) = 2x^3 - 3$ and $g(x) = 3x^2 + x - 2$.

Which of the following is $f(x) \times g(x)$?

A $5x^5 + 2x^4 - 4x^3 - 9x^2 - 3x + 6$

B $6x^5 + x^4 - 2x^3 - 9x^2 - 3x + 6$

C $6x^5 + 2x^4 - 4x^3 + 9x^2 + 3x - 6$

D $6x^5 + 2x^4 - 4x^3 - 9x^2 - 3x + 6$

E $6x^5 + 2x^4 - 4x^3 - 9x^2 - x - 2$

3 Which of the following is $(3x - 2)(x + 1) - x(1 - 2x)$?

A $5x^2 - 2$ B $5x^2 - 3x - 1$ C $5x^2 + 2$ D $3x^2 - 2x - 2$ E $x^2 - 2$

4 Which of the following graphs represents the shape of the curve $y = -x^4 + 5x^3 - 5x^2 - 5x + 6$?

A

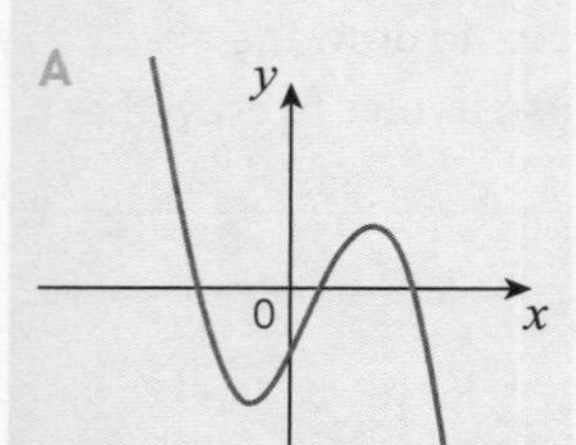

B

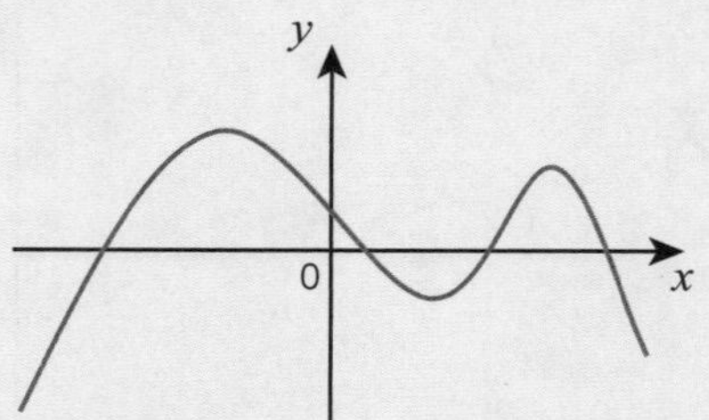

C

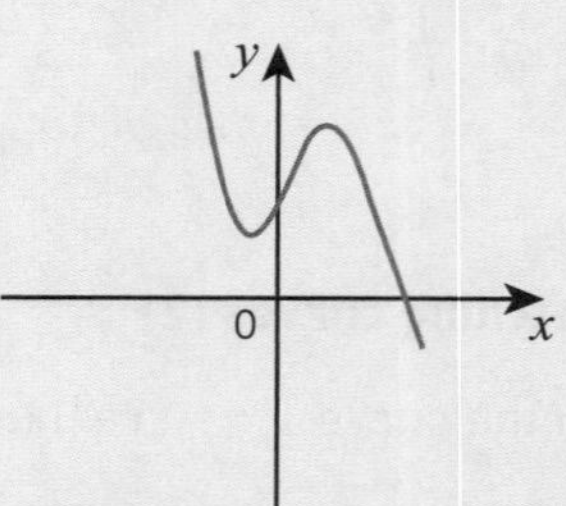

D

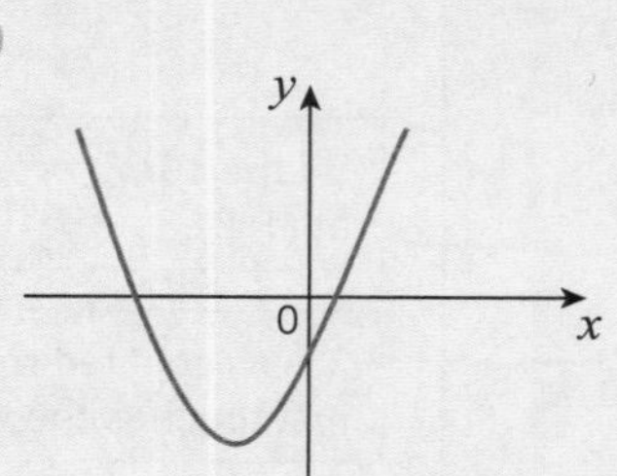

E

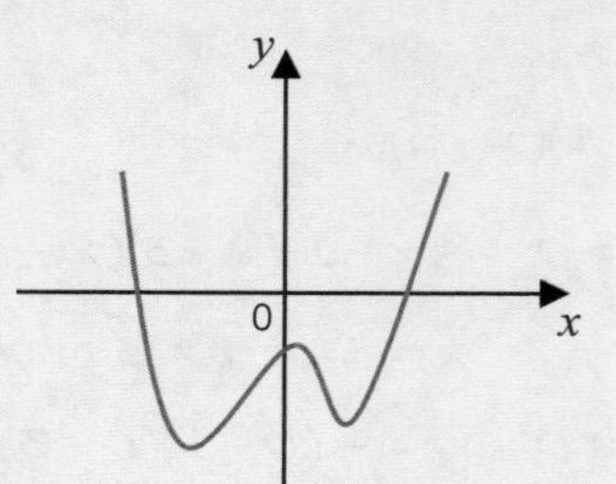

5 Which of the following is the graph of the curve $y = (x^2 - 1)(x + 1)$?

A

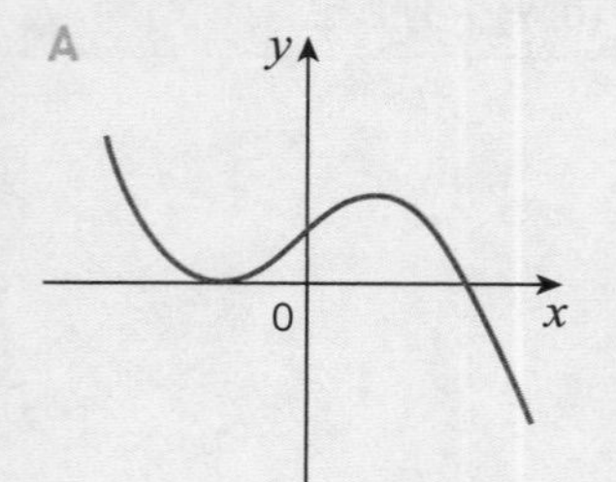

B

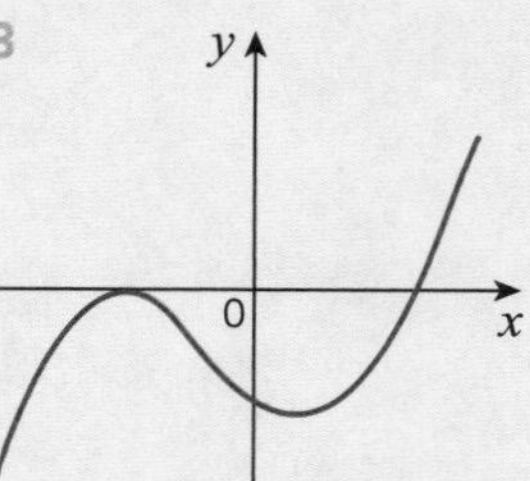

C

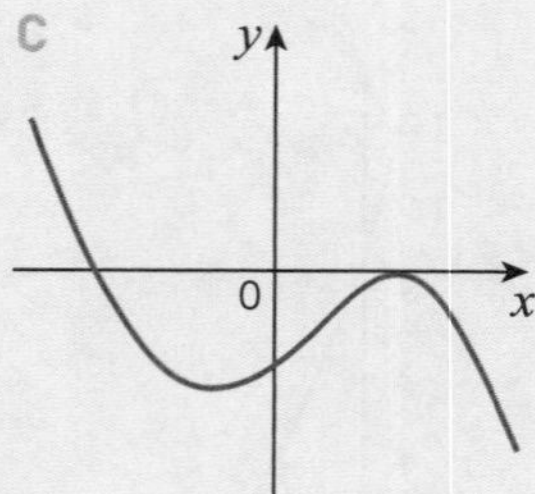

D

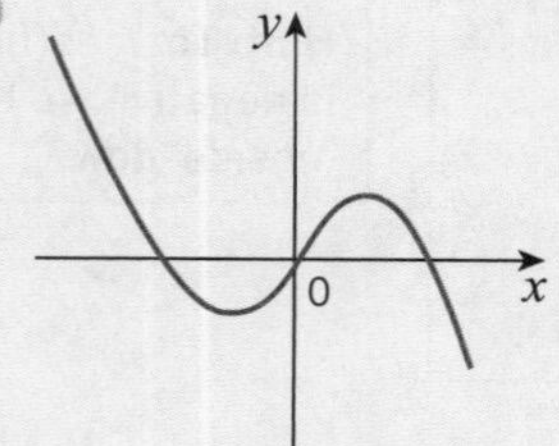

E

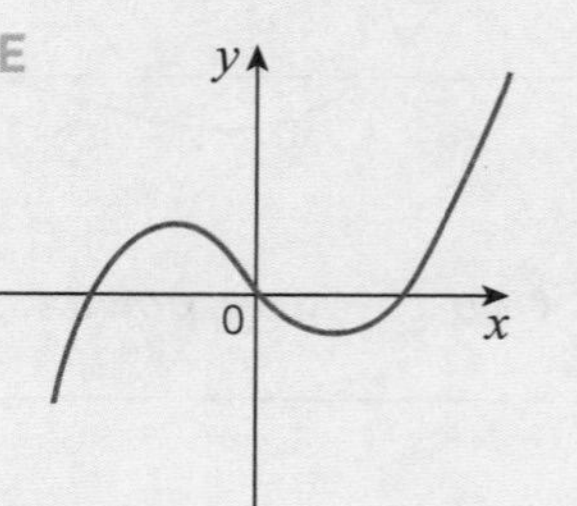

Full worked solutions online

CHECKED ANSWERS

Exam-style question

You are given that $y = (3 - x)(x - 1)^2$.

i Show that $y = -x^3 + 5x^2 - 7x + 3$.

ii Sketch the graph of the polynomial.

Short answers on page 154

Full worked solutions online

CHECKED ANSWERS

Dividing polynomials

REVISED

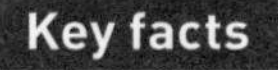

Key facts

1 The words you use for dividing two whole numbers are shown below.

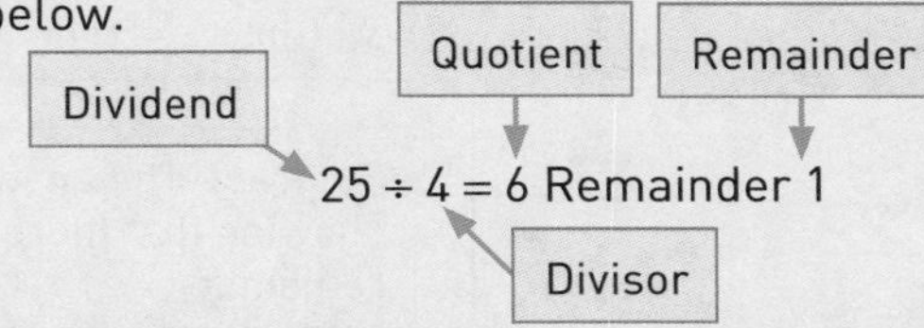

2 One way of dividing polynomials is to set them out in columns, rather like a long division using numbers. See example 1.
Reminder: long division, using numbers:

325 ÷ 25

$$\begin{array}{r} 13 \\ 25\,\overline{)\,325} \\ -25\downarrow \\ \hline 75 \\ -\ 75 \\ \hline 0 \end{array}$$

- 1×25
- 3×25
- No remainder.

3 Another way (when there is no remainder) is to set the question out as a multiplication and then compare the coefficients to find the quotient. See example 2.

Worked examples

1 Using long division to find a quotient

Divide $2x^3 - 3x^2 + 6x - 5$ by $x - 1$.

Hint: Lay it out in columns.

Solution

$$\begin{array}{r} 2x^2 \qquad\qquad\quad \\ x-1\,\overline{)\,2x^3 - 3x^2 + 6x - 5} \\ -\ \ 2x^3 - 2x^2 \qquad\qquad\ \ \\ \hline -x^2 + 6x \qquad\ \ \end{array}$$

Divide $2x^3$ by x to get $2x^2$.

Multiply $(x - 1)$ by $2x^2$.

Subtract $2x^3 - 2x^2$ from $2x^3 - 3x^2$ which gives $-x^2$.

Bring down $6x$.

$$\begin{array}{r} 2x^2 - x \\ x-1\overline{)2x^3 - 3x^2 + 6x - 5} \\ -\quad 2x^3 - 2x^2 \\ \hline -x^2 + 6x \\ -\quad -x^2 + x \\ \hline +5x - 5 \end{array}$$

Divide $-x^2$ by x to get $-x$ and write the answer here.

Multiply $(x - 1)$ by $-x$.

Subtract which gives $5x$.

Bring down -5.

$$\begin{array}{r} 2x^2 - x + 5 \\ x-1\overline{)2x^3 - 3x^2 + 6x - 5} \\ -\quad 2x^3 - 2x^2 \\ \hline -x^2 + 6x \\ -\quad -x^2 + x \\ \hline +5x - 5 \\ -\quad +5x - 5 \\ \hline 0 \end{array}$$

Divide $5x$ by x and write the answer here.

Multiply $(x - 1)$ by $+5$.

Subtract. The answer is zero showing that there is no remainder.

So $(2x^3 - 3x^2 + 6x - 5) \div (x - 1) = 2x^2 - x + 5$

2 Comparing coefficients to find a quotient

Divide $2x^3 - 3x^2 + 6x - 5$ by $x - 1$.

Hint: The method that follows is suitable only when the remainder is 0.

Solution

$2x^3 - 3x^2 + 6x - 5 = (x - 1)(ax^2 + bx + c)$

Step 1: Look at the term in x^3: $2x^3 = x \times ax^2 \Rightarrow a = 2$

So $\quad 2x^3 - 3x^2 + 6x - 5 = (x - 1)(2x^2 + bx + c)$

Since $2x^3 \div x = 2x^2$ the first term in the right bracket should be $2x^2$.

Step 2: Look at the constant term: $-5 = -1 \times c \Rightarrow c = 5$

So $\quad 2x^3 - 3x^2 + 6x - 5 = (x - 1)(2x^2 + bx + 5)$

The constant term on the left is -5.

To get -5 when you multiply out the brackets you need '+5' in the last bracket.

Step 3: Look at the term in x: $+6x = x \times 5 + (-1) \times bx$

$\Rightarrow \quad 6x = 5x - bx$

Look for the pairs of terms that multiply together to give a term in x

Comparing coefficients of x gives: $6 = 5 - b \Rightarrow b = -1$

$\Rightarrow \quad 2x^3 - 3x^2 + 6x - 5 = (x - 1)(2x^2 - x + 5)$

On the left side you have $6x$ and on the right you have $(5 - b)x$.

Alternative Step 3:

Look at the coefficients of x^2 to find b:

$-3x^2 = x \times bx + (-1) \times 2x^2 \Rightarrow -3x^2 = bx^2 - 2x^2$

Look for the pairs of terms that multiply together to give a term in x^2.

Comparing coefficients of x^2 gives: $-3 = b - 2 \Rightarrow b = -1$

On the left side you have $-3x^2$ and on the right you have $(b - 2)x^2$.

$\Rightarrow 2x^3 - 3x^2 + 6x - 5 = (x - 1)(2x^2 - x + 5)$ as before.

Hint: Check your result by multiplying out the brackets on the right-hand side.

$(x-1)(2x^2-x+5)$

$=x(2x^2-x+5)-1\times(2x^2-x+5)$

$=2x^3-x^2+5x-2x^2+x-5$

$=2x^3-3x^2+6x-5$ ✓

Test yourself

TESTED

1 When $(3x^4-2x^3-x+1)$ is divided by $(x-1)$ the answer is $(3x^3+x^2+x)$ and the remainder is +1. Four of the statements below are true and one is false. Which one is false?

A $3x^4-2x^3-x+1=(x-1)(3x^3+x^2+x)+1$.

B The quotient is a polynomial of order 3.

C The remainder is a polynomial of order 1.

D There is no remainder when $3x^4-2x^3-x$ is divided by $x-1$.

E The remainder is a constant.

2 The quotient when $3x^3-5x^2+6x-4$ is divided by $x-1$ is

A $3x^2+2x+4$ B $-3x^2-2x+4$ C $3x^2-2x-4$ D $3x^2-2x+4$ E $3x^3-1$

3 You are given that $2x^3+x^2-7x-6=(x-2)(2x^2+bx+3)$. The value of b is:

A -3 B 5 C 3 D -5 E $-\frac{1}{4}$

4 You are given that $6x^3+13x^2-2x-12=(2x+3)(ax^2+bx+c)$.
The quadratic factor (ax^2+bx+c) is

A $3x^2+2x+4$ B $3x^2+2x-4$ C $3x^2-\frac{10}{3}x+4$ D $3x^2+\frac{10}{3}x-4$ E $3x^2-2x-4$

5 When $2x^3-x^2-x-1$ is divided by $x-2$ the remainder is:

A -23 B 9 C -11 D -15 E 13

Full worked solutions online

CHECKED ANSWERS

Exam-style question

The polynomial x^4-3x^2+3x+d is divisible by $(x+1)$.

Divide x^4-3x^2+3x+d by $(x+1)$ and hence find the value of d.

Short answers on page 154

Full worked solutions online

CHECKED ANSWERS

The factor theorem

REVISED

Key facts

1 The factor theorem says:
If $(x - a)$ is a factor of $f(x)$ then $f(a) = 0$ and $x = a$ is a root of the equation $f(x) = 0$.

Conversely, if $f(a) = 0$ then $(x - a)$ is a factor of $f(x)$.

2 You can use the factor theorem to test for factors or roots of a polynomial.
For example, to find factors of $f(x) = 2x^3 + x^2 - 13x + 6$, look at the constant term '+6'
6 is divisible by: ±1, ±2, ±3, ±6
$f(1) = 2 \times 1^3 + 1^2 - 13 \times 1 + 6 = 2 + 1 - 13 + 6 = -4$ ✗
So $x = 1$ is **not** a root and $(x - 1)$ is **not** a factor.
$f(2) = 2 \times 2^3 + 2^2 - 13 \times 2 + 6 = 16 + 4 - 26 + 6 = 0$ ✓
$\Rightarrow x = 2$ is a root and $(x - 2)$ is a factor of $f(x) = 2x^3 + x^2 - 13x + 6$.

3 Once you have used the factor theorem to find one (or more) factors you can use long division to factorise the polynomial fully. You can then find the roots of the polynomial and sketch its graph.

Hint: When $(ax + b)$ is a factor of a polynomial $f(x)$ then $f\left(-\frac{b}{a}\right) = 0$ and $x = -\frac{b}{a}$ is a root of $f(x)$.

To find $f(a)$ substitute $x = a$ into the polynomial.
For example, given $f(x) = 3x - 1$ then $f(2) = 3 \times 2 - 1 = 5$.

Any integer roots will be factors of the constants term.

Worked examples

1 Showing a linear expression is a factor

Show that $(x + 1)$ is a factor of $f(x) = x^3 + 2x^2 - 5x - 6$.

Solution

Using the factor theorem:

If $(x + 1)$ is a factor of $f(x)$ then $f(-1) = 0$.

$$f(-1) = (-1)^3 + 2 \times (-1)^2 - 5 \times (-1) - 6$$
$$= -1 + 2 + 5 - 6$$
$$= 0$$

So $(x + 1)$ is a factor of $f(x) = x^3 + 2x^2 - 5x - 6$.

Hint: When the question says 'show', you must show **all** of your working.

2 Using the factor theorem to solve equations (1)

You are given $f(x) = x^3 + 3x^2 - 4x - 12$.

f(x) is a cubic so it will have at most three factors.

i Find the values of $f(1)$, $f(-1)$, $f(2)$, $f(-2)$ and state two factors of the polynomial.

ii Hence find two roots of the equation $x^3 + 3x^2 - 4x - 12 = 0$.

iii Use the factor theorem to decide whether the third root is $x = 3$ or $x = -3$.

Solution

i $f(1) = 1^3 + 3 \times 1^2 - 4 \times 1 - 12 = -12$
$\Rightarrow (x-1)$ is not a factor.

$f(-1) = (-1)^3 + 3 \times (-1)^2 - 4 \times (-1) - 12 = -6$
$\Rightarrow (x+1)$ is not a factor.

$f(2) = 2^3 + 3 \times 2^2 - 4 \times 2 - 12 = 0$
$\Rightarrow (x-2)$ is a factor.

When $f(a) = 0$ then $(x-a)$ is a factor.

$f(-2) = (-2)^3 + 3 \times (-2)^2 - 4 \times (-2) - 12 = 0$
$\Rightarrow (x+2)$ is a factor.

So two factors of $f(x)$ are $(x-2)$ and $(x+2)$.

ii $x = 2$ and $x = -2$ are roots of $f(x) = 0$.

$f(a) = 0 \Rightarrow x = a$ is a root of $f(x) = 0$.

iii $f(3) = 3^3 + 3 \times 3^2 - 4 \times 3 - 12 = 54 - 24 = 30$ ✗

$f(-3) = (-3)^3 + 3 \times (-3)^2 - 4 \times (-3) - 12 = 0$ ✓

So $x = -3$ is a root of $x^3 + 3x^2 - 4x - 12 = 0$.

So $(x+3)$ is the third factor.
Factorising fully gives $f(x) = (x-2)(x+2)(x+3)$.

3 Using the factor theorem to solve equations (2)

Given that $f(x) = x^3 - x^2 - 4x + 4$

i factorise $f(x)$ fully;

ii solve the equation $f(x) = 0$.

Hint: Factorise completely or **factorise fully** means keep going until there are no further factors to be found.

$f(a) = 0 \Rightarrow x = a$ is a root of $f(x) = 0$.

Solution

i $f(x) = x^3 - x^2 - 4x + 4$

The constant term is 4 and 4 is divisible by 1, −1, 2, −2, 4 and −4.

Hint: To find any integer roots of the equation look at the constant term of the polynomial.

$f(1) = 1^3 - 1^2 - 4 \times 1 + 4 = 0$ ✓ $\Rightarrow (x-1)$ is a factor.

$f(-1) = (-1)^3 - (-1)^2 - 4 \times (-1) + 4 = 6$ ✗

$f(2) = 2^3 - 2^2 - 4 \times 2 + 4 = 0$ ✓ $\Rightarrow (x-2)$ is a factor.

$f(-2) = (-2)^3 - (-2)^2 - 4 \times (-2) + 4 = 0$ ✓ $\Rightarrow (x+2)$ is a factor.

You have found all three factors so you don't need to check $f(4)$ and $f(-4)$.

$\Rightarrow f(x) = x^3 - x^2 - 4x + 4 = (x-1)(x-2)(x+2)$

ii The solution is $x = 1$, $x = 2$ or $x = -2$.

4 Using the factor theorem to sketch a curve

Given that $(x+1)$ is a factor of $f(x) = x^3 - 3x^2 - x + 3$

i factorise $x^3 - 3x^2 - x + 3$ completely

ii solve the equation $x^3 - 3x^2 - x + 3 = 0$

iii sketch the curve of $y = x^3 - 3x^2 - x + 3$.

Solution

i $(x+1)$ is a factor of $f(x) = x^3 - 3x^2 - x + 3$

$$
\begin{array}{r}
x^2 - 4x + 3 \\
x+1 \,\big)\, x^3 - 3x^2 - x + 3 \\
- \quad x^3 + x^2 \qquad\qquad \\
\hline
-4x^2 - x \qquad \\
- \quad -4x^2 - 4x \qquad \\
\hline
3x + 3 \\
- \quad 3x + 3 \\
\hline
0
\end{array}
$$

To factorise completely, divide $x^3 - 3x^2 - x + 3$ by $(x + 1)$.

Factorising $x^2 - 4x + 3$ gives $(x - 1)(x - 3)$.

So $x^3 - 3x^2 - x + 3 = (x+1)(x^2 - 4x + 3)$

$= (x + 1)(x - 1)(x - 3)$

ii Solve $x^3 - 3x^2 - x + 3 = 0$

or $(x+1)(x-1)(x-3) = 0$

So either $(x+1) = 0 \Rightarrow x = -1$

or $(x-1) = 0 \Rightarrow x = 1$

or $(x-3) = 0 \Rightarrow x = 3$.

So the solution to the equation is $x = -1$, 1, or 3.

iii The curve $y = x^3 - 3x^2 - x + 3$ crosses the x-axis at $(-1, 0)$, $(1, 0)$ and $(3, 0)$.

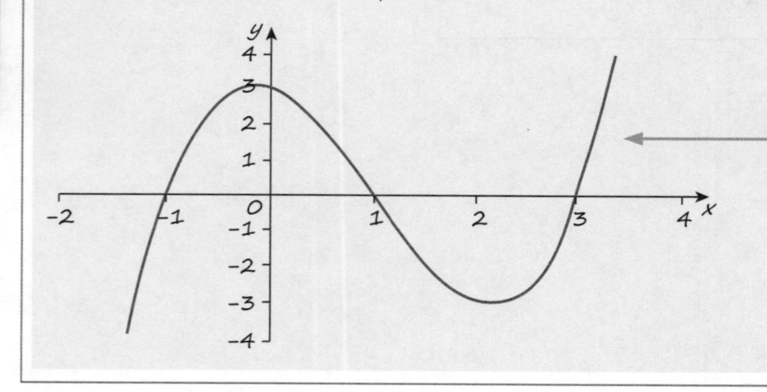

When $x = 0$ then $f(x) = 3$ so the curve cuts the y-axis at $(0, 3)$.

Test yourself

TESTED

1 $(x-2)$ is a factor of one of these polynomials. Which one is it?

A $x^3 - 2x^2 - 3x + 1$ B $x^4 - 2x^3$ C $2x^3 - x^2 - 4$

D $x^2 + 4x - 16$ E $2x^3 - 4x^2 - 3x + 2$

2 You are given $f(x) = 3x^3 - 2x^2 - 3x + 2$

Four of the statements below are true and one is false. Which one is false?

A $(x-1)$ is a factor of the polynomial $f(x)$.

B When the polynomial $3x^3 - 2x^2 - 3x + 2$ is divided by $(x+1)$ there is no remainder.

C $(x+2)$ is not a factor of the polynomial $f(x)$.

D $(3x+2)$ is a factor of the polynomial $f(x)$.

E There are three different values of x for which $f(x) = 0$.

3 A polynomial $f(x)$ is given by $x^3 - 3x^2 - 7x - 15$. Which one of these is a factor of $f(x)$?

A $(x-1)$ B $(x-3)$ C $(x-5)$ D $(x+1)$ E $(x+3)$

4 A polynomial is given by $f(x) = x^4 + 2x^3 + 3x^2 + 4x + 2$.

Four of the following statements are false and one is true. Which one is true?

A The four roots of $x^4 + 2x^3 + 3x^2 + 4x + 2 = 0$ are $x = 1$, $x = -1$, $x = 2$ and $x = -2$.

B There are two different roots.

C The curve $y = f(x)$ touches the x-axis.

D $(x+1)$ is a factor of $f(x)$ but $(x+1)^2$ is not a factor.

E $(x+2)$ is a factor of $f(x)$.

5 Which of the following is the equation of this curve?

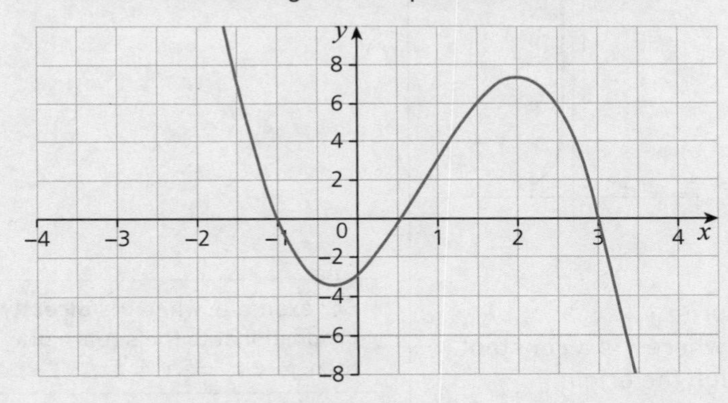

A $y = 2x^4 + x^3 - 19x^2 - 9x + 9$ B $y = 2x^3 - 5x^2 - 4x + 3$ C $y = x^2 - 2x + 3$

D $y = -2x^2 - x + 1$ E $y = -2x^3 + 5x^2 + 4x - 3$

Full worked solutions online

CHECKED ANSWERS

Exam-style question

Given that $(x-1)$ and $(x+3)$ are factors of $x^3 - x^2 + ax + b$, find the values of a and b.

Short answers on page 154

Full worked solutions online

CHECKED ANSWERS

Chapter 8 Graphs and transformations

About this topic

This topic gives further techniques that help when setting up equations, sketching graphs and solving problems.

Being able to start with a simple curve and transform it to derive the shape of a curve with a more complicated equation is a very important mathematical skill. It is often used when the form of the curve is trigonometrical, particularly the sine wave.

Before you start, remember

- how to complete the square for a quadratic function
- polynomial functions and their graphs
- how to sketch $y = \sin x$, $y = \cos x$ and $y = \tan x$.

Curve sketching and transformations

REVISED

Key facts

1 Make sure you know the shapes of these common curves:

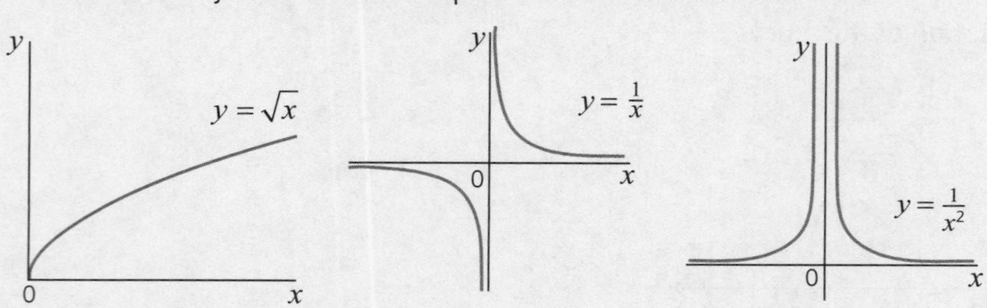

2 The symbol $\propto$ means 'proportional to'.

- When y is **directly proportional** to x you write $y \propto x$.
  You can write this as an equation: $y = kx$ where k is a constant.
  The graph of $y = kx$ is a straight line through the origin.
- When y is **inversely proportional** to x you write $y \propto \frac{1}{x}$
  You can write this as an equation: $y = \frac{k}{x}$ where k is a constant.
  You can use given values of y and x to find the value of k.

For example, when y is **directly** proportional to the square of x then $y \propto x^2 \Rightarrow y = kx^2$.

For example, when y is inversely proportional to the cube root of x then $y \propto \frac{1}{\sqrt[3]{x}} \Rightarrow y = \frac{k}{\sqrt[3]{x}}$

3 Transformations of $y = f(x)$ are:

Translations (for positive a and b)

- parallel to x-axis through $\begin{pmatrix} a \\ 0 \end{pmatrix}$

  $y = f(x - a)$

- parallel to y-axis through $\begin{pmatrix} 0 \\ b \end{pmatrix}$

  $y = f(x) + b$

Stretches:

- parallel to x-axis scale factor $\frac{1}{a}$

  $y = f(ax)$

- parallel to y-axis scale factor a

  $y = af(x)$

Reflections:

- in x-axis

  $y = -f(x)$

- in y-axis

  $y = f(-x)$

Hint: $y = f(x + a)$ translates the curve a units left.

$y = f(x - b)$ translates the curve b units down.

a units right.

b units up.

A reflection is just a special case of a stretch with scale factor –1.

Worked examples

1 Using proportion

A coin is dropped down a well.

The time taken, t seconds, for the coin to reach the bottom of the well is directly proportional to the square root of the depth of the well, d metres.

A coin takes 2 seconds to reach the bottom of a well that is 20 m deep.

i Find the equation connecting t and d

ii draw the graph of

a t against d

b t against $\sqrt{d}$.

Solution

i $t \propto \sqrt{d} \Rightarrow t = k\sqrt{d}$

When $d = 20$, $t = 2 \Rightarrow 2 = k\sqrt{20} \Rightarrow k = \frac{2}{\sqrt{20}} = \frac{2}{\sqrt{4 \times 5}}$

$= \frac{1}{\sqrt{5}} = \frac{\sqrt{5}}{\sqrt{5}}$

So $t = \frac{\sqrt{5}\sqrt{d}}{5} = \frac{\sqrt{5d}}{5}$

Substitute in the values of t and d and solve to find k.

ii a

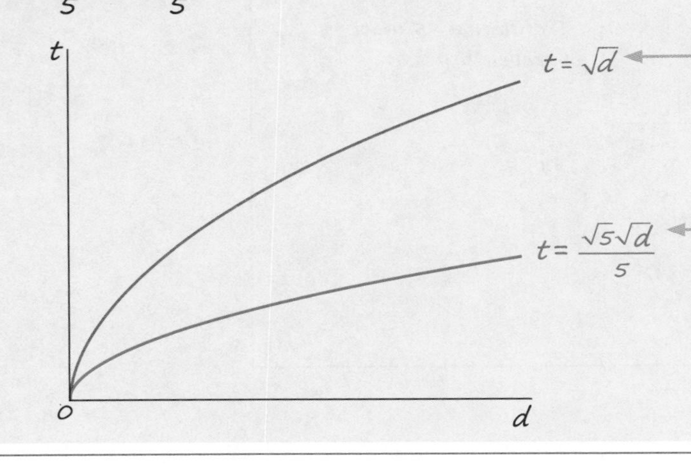

Start by sketching the graph of $t = \sqrt{d}$.

The graph of $t = \frac{\sqrt{5}\sqrt{d}}{5}$ is a one way stretch scale factor $\frac{\sqrt{5}}{5} = \sqrt{5}$ parallel to the y-axis.

b

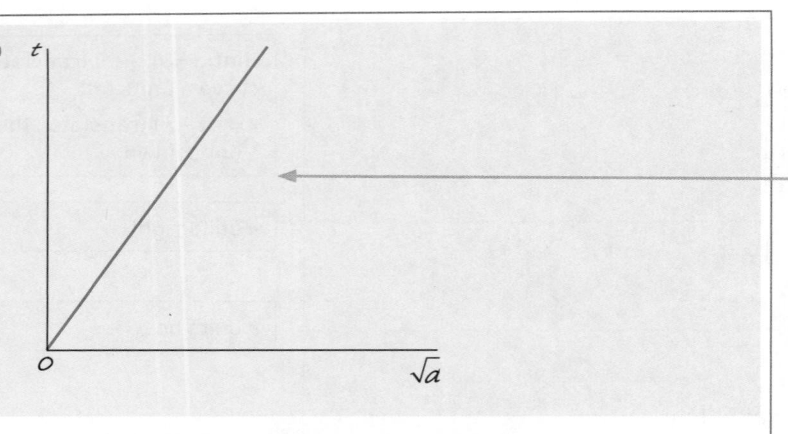

t is directly proportional to $\sqrt{d}$ so the graph of t against $\sqrt{d}$ is a straight line through the origin with gradient $k = \frac{\sqrt{5}}{5}$.

2 Using transformations to sketch curves

Starting with the graph of $y = x^2$, sketch the graphs of

i $y = (x - 4)^2$ ii $y = (x^2 - 5)$ iii $y = (x - 4)^2 - 5$.

Notice that this a quadratic in 'completed square' form.

In each case describe the transformation.

Solution

i

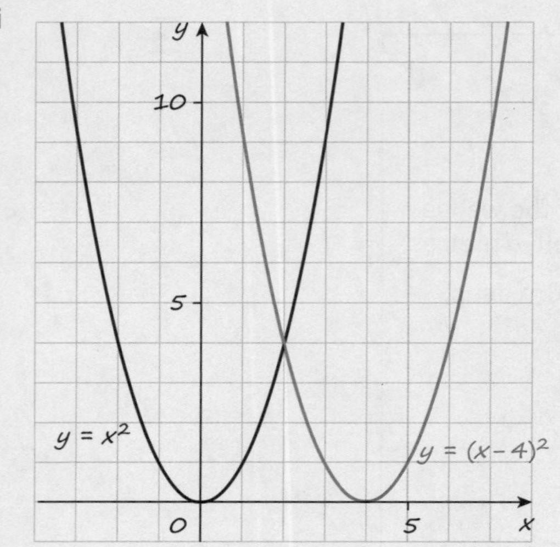

Translation $\begin{pmatrix} 4 \\ 0 \end{pmatrix}$

Translation +4 units parallel to x-axis

ii

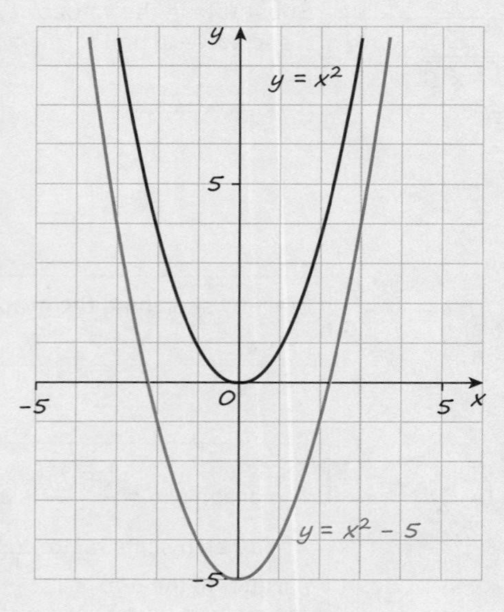

Translation $\begin{pmatrix} 0 \\ -5 \end{pmatrix}$

Translation −5 units parallel to y-axis

iii

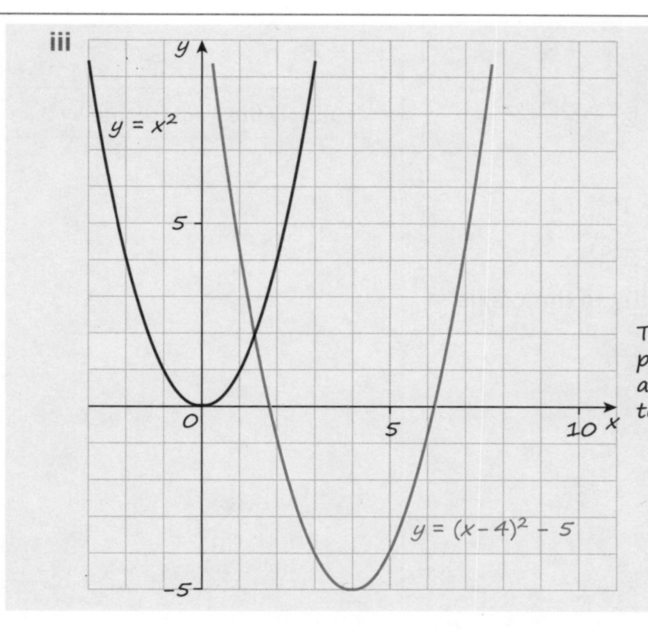

Translation 4 units parallel to x-axis and 5 units parallel to y-axis.

Translation $\begin{pmatrix} 4 \\ -5 \end{pmatrix}$

3 Applying transformations to any function

The diagram shows the graph of $y = \mathrm{f}(x)$.

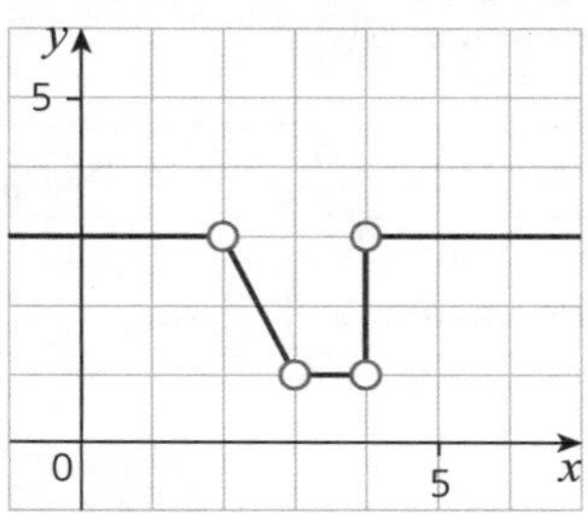

i Write the equation of the graph given in this diagram.

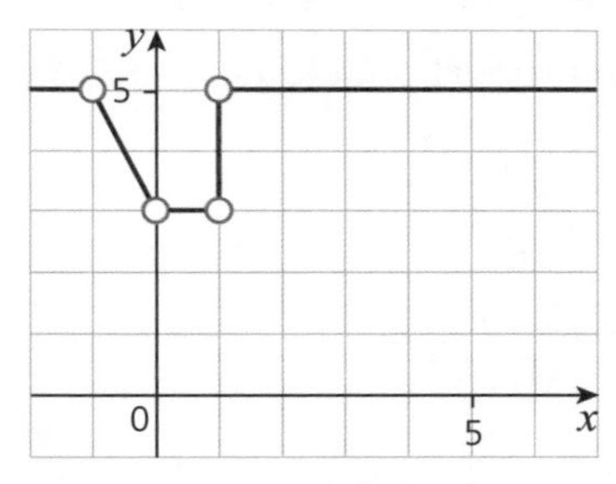

ii Sketch the graph of $y = \mathrm{f}(-x)$.

iii Write down the coordinates of the point where $y = 2\mathrm{f}(x)$ intersects the y-axis.

Solution

i *The graph has been translated through* $\begin{pmatrix} -3 \\ 2 \end{pmatrix}$ *so the new equation is* $y = \mathrm{f}(x + 3) + 2$.

ii

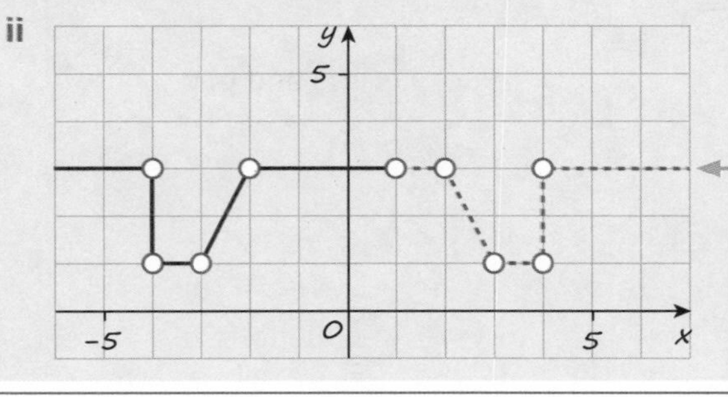

Reflect $y = \mathrm{f}(x)$ in the y-axis to get $y = \mathrm{f}(-x)$

Common mistake: Take care with your signs:

$\mathrm{f}(x$ **+ 3**) represents a translation of **3 units to the left**.

$\mathrm{f}(x$ **− 3**) represents a translation of **3 units to the right**.

iii $y = f(x)$ intersects the y-axis at $(0, 3)$

So $y = 2f(x)$ intersects the y-axis at $(0, 6)$

Multiply the y coordinate by 2.

4 Finding the equation of a transformed curve

A cubic function $f(x)$ is given by $f(x) = (x + 2)(2x + 1)(x - 3)$.

The curve $y = f(x)$ is stretched by scale factor $\frac{1}{2}$ parallel to the x-axis.

The equation of the stretched curve is $y = g(x)$.

i Find the equation of $g(x)$ in terms of x.

The curve $y = f(x)$ is translated through $\begin{pmatrix} 3 \\ 0 \end{pmatrix}$.

The equation of the translated curve is $y = h(x)$.

ii Find the equation of $h(x)$ in terms of x.

Solution

i $g(x) = f(2x)$

$= (2x + 2)(2(2x) + 1)(2x - 3)$

$= (2x + 2)(4x + 1)(2x - 3)$

Stretch scale factor $\frac{1}{2}$ parallel to the x-axis.

Replace 'x' with '$2x$'.

ii $h(x) = f(x - 3)$

$= ((x - 3) + 2)(2(x - 3) + 1)((x - 3) - 3)$

$= (x - 1)(2x - 5)(x - 6)$

Translation of 3 units to the right.

Replace 'x' with '$x - 3$'.

Test yourself

TESTED

1 Which of these graphs show that

a y is directly proportional to x

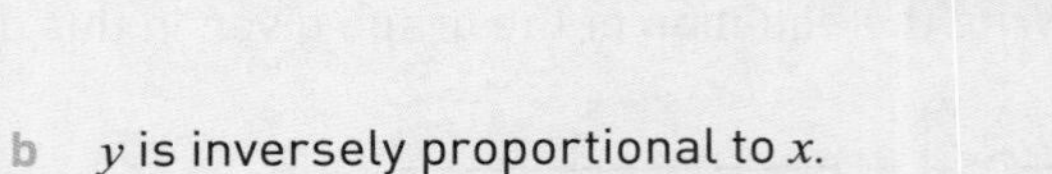

b y is inversely proportional to x.

i

ii

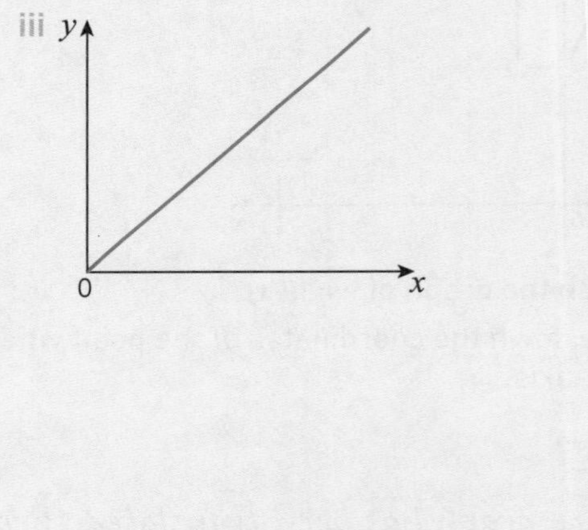

iii

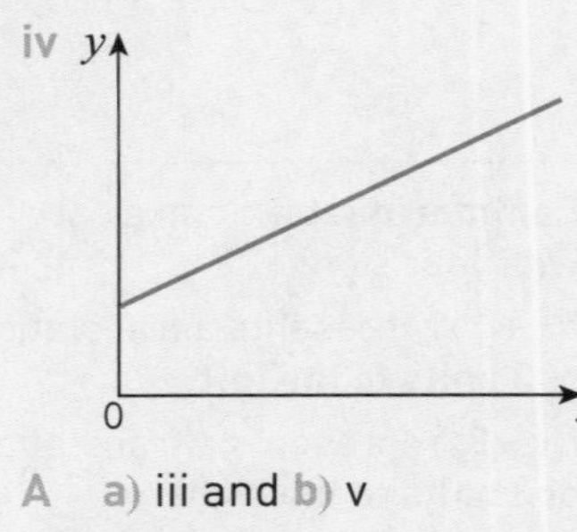

iv

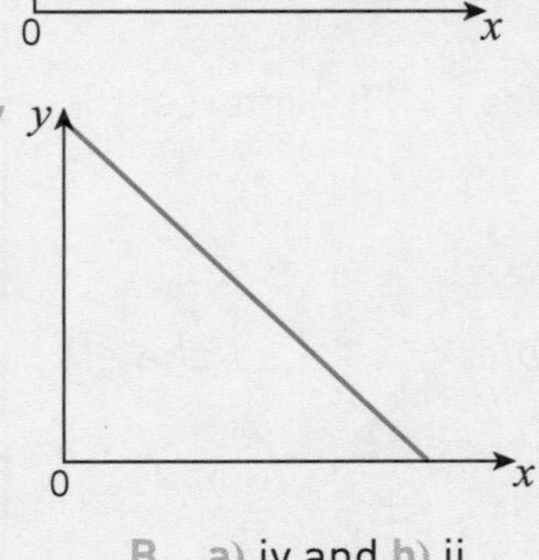

v

A a) iii and b) v

B a) iv and b) ii

C a) i and b) v

D a) i and b) ii

E a) iii and b) ii

For questions 2 and 3 you will need to use the diagram below.

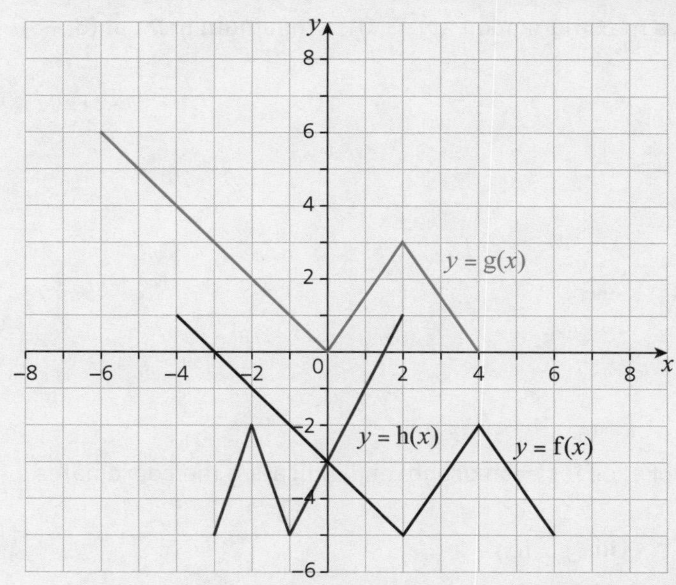

2 The diagram shows the graph of $y = \mathrm{f}(x)$ and its image $y = \mathrm{g}(x)$ after a transformation. The equation of the image is:

A $y = \mathrm{f}(x + 2) - 5$

B $y = \mathrm{f}(x - 2) + 5$

C $y = \mathrm{f}(x - 2) - 5$

D $y = \mathrm{f}(x + 2) + 5$

E $y = \mathrm{f}(x + 5) + 2$

3 The diagram shows the graph of $y = \mathrm{f}(x)$ and its image $y = \mathrm{h}(x)$ after a transformation. The equation of the image is:

A $y = \mathrm{f}(2x)$

B $y = \mathrm{f}\left(\frac{1}{2}x\right)$

C $y = \mathrm{f}(-2x)$

D $y = \mathrm{f}\left(-\frac{1}{2}x\right)$

E $y = -\frac{1}{2}\mathrm{f}(x)$

4 The curve $y = x^2 - 4x$ is translated and the equation of the new curve is $y = (x - 1)^2 - 4(x - 1) + 2$. What are the coordinates of the vertex of the new curve?

A $(1, 2)$

B $(1, -2)$

C $(3, -2)$

D $(1, 5)$

E $(1, -6)$

5 The curve $y = x^2 - 2x + 3$ is translated through $\begin{pmatrix} 2 \\ -4 \end{pmatrix}$. Find the equation of the new curve.

A $y = x^2 - 2x - 1$

B $y = x^2 - 6x + 11$

C $y = x^2 - 6x + 15$

D $y = 4x^2 - 4x - 1$

E $y = x^2 - 6x + 7$

Full worked solutions online

CHECKED ANSWERS

Exam-style question

The diagram shows the graph of $y = f(x)$ which has a maximum point at $(-3, 3)$, a minimum point at $(3, -3)$, and passes through the origin.

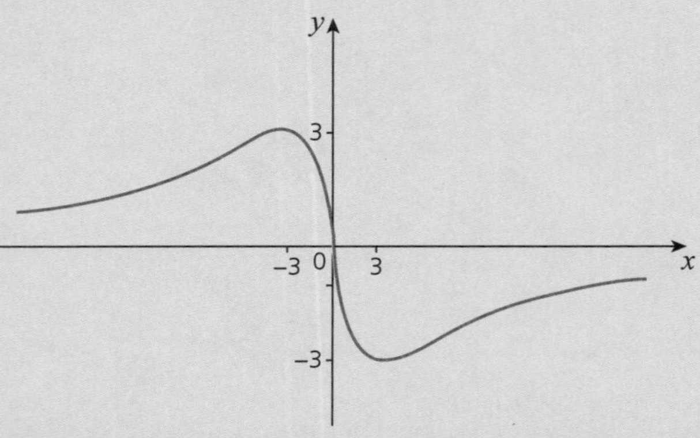

Sketch the following graphs, using a separate set of axes for each graph, and indicating the coordinates of the turning points.

i $y = 2f(x)$ ii $y = f(2x)$ iii $y = f(x) + 2$

Short answers on page 154

Full worked solutions online

CHECKED ANSWERS

Transformations and graphs of trigonometric functions

REVISED

Key fact

You can apply the transformations given on page 73 to trigonometric functions.

Worked examples

1 Using transformations

Starting with the curve $y = \sin\theta$, describe and show how transformations can be used to sketch these curves:

i $y = \sin\theta + 2$ ii $y = 3\sin\theta$ iii $y = -\sin\theta$

iv $y = \sin(30° + \theta)$ v $y = \sin\left(\frac{\theta}{2}\right)$.

Solution

i The curve $y = \sin\theta + 2$ is obtained from the curve $y = \sin\theta$ by a translation $\begin{pmatrix} 0 \\ 2 \end{pmatrix}$.

Hint: Remember, to obtain $y = f(\theta) + a$ from $y = f(\theta)$ you translate $y = f(\theta)$ through a units vertically.

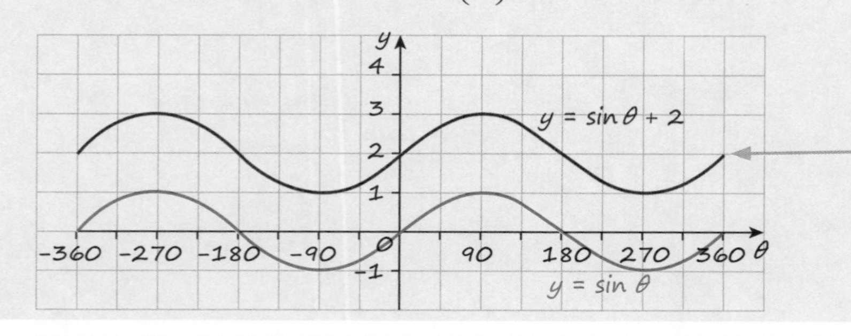

The curve $y = \sin\theta + 2$ oscillates between a minimum of $y = 1$ and a maximum of $y = 3$.

ii The curve $y = 3\sin\theta$ is obtained from the curve $y = \sin\theta$ by a stretch with scale factor 3, parallel to the y-axis.

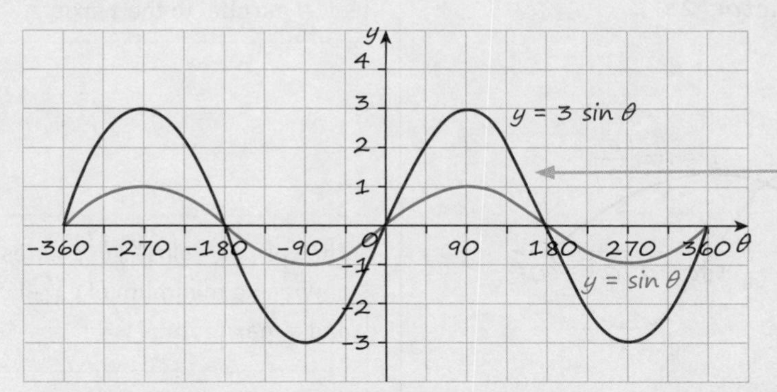

Hint: Remember, to obtain $y = a\text{f}(\theta)$ from $y = \text{f}(\theta)$ you stretch $y = \text{f}(\theta)$ by a scale factor of a parallel to the y-axis.

The curve $y = 3\sin\theta$ oscillates between a minimum of $y = -3$ and a maximum of $y = 3$.

iii The curve $y = -\sin\theta$ is obtained from the curve $y = -\sin\theta$ by reflection in the x-axis.

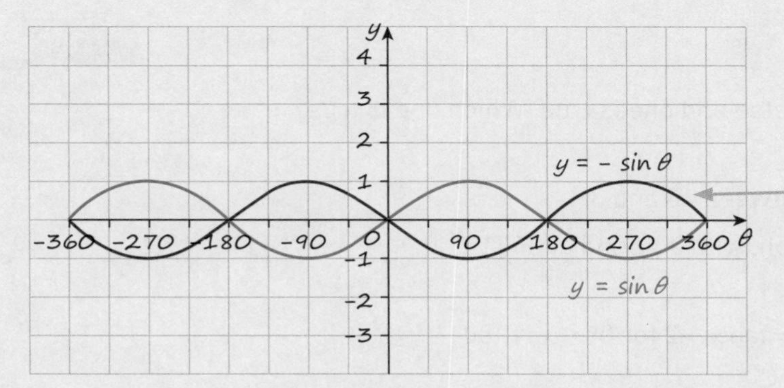

Hint: Remember, to obtain $y = -\text{f}(\theta)$ from $y = \text{f}(\theta)$ you reflect $y = \text{f}(\theta)$ in the x-axis.

The curve $y = -\sin\theta$ oscillates between a minimum of $y = -1$ and a maximum of $y = 1$.

iv The curve $y = \sin(\theta + 30°)$ is obtained from the curve $y = \sin\theta$ by translation $\begin{pmatrix} -30° \\ 0 \end{pmatrix}$ or 30° to the left along the x-axis.

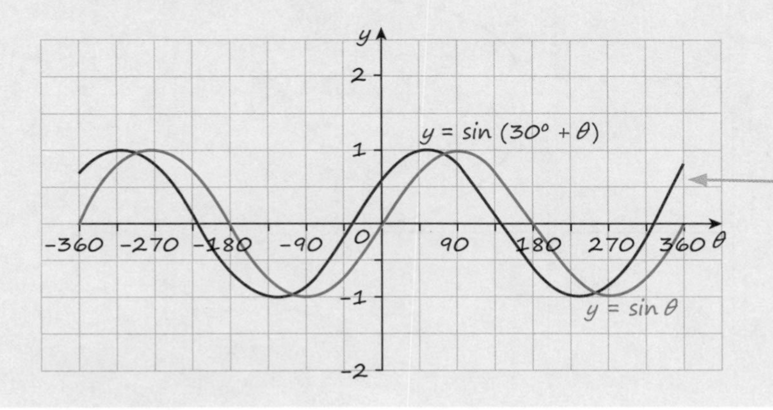

Hint: Remember, to obtain $y = \text{f}(\theta + a)$ from $y = \text{f}(\theta)$ you translate $y = \text{f}(\theta)$ a units to the left.

The curve $y = \sin(\theta + 30°)$ oscillates between a minimum of $y = -1$ and a maximum of $y = 1$.

v The curve $y = \sin\left(\frac{\theta}{2}\right)$ is obtained from the curve $y = \sin\theta$ by a stretch of scale factor 2.

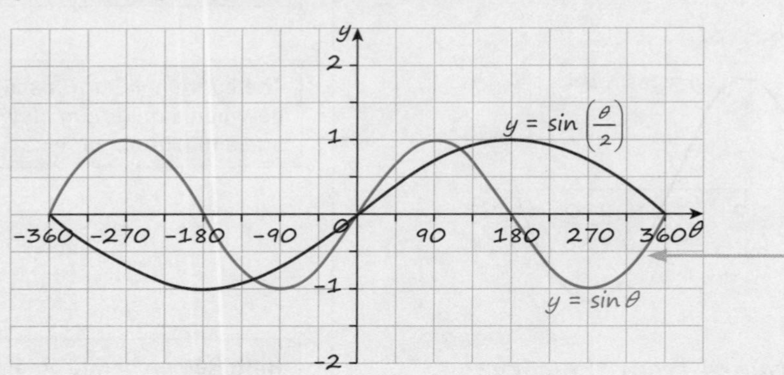

Hint: Remember, to obtain $y = f(a\theta)$ from $y = f(\theta)$ you stretch $y = f(\theta)$ by a scale factor of $\frac{1}{a}$ parallel to the x-axis.

The curve $y = \sin\left(\frac{\theta}{2}\right)$ oscillates between a minimum of $y = -1$ and a maximum of $y = 1$.

Test yourself

TESTED

1 Three of the following statements are false and one is true. Which one is true?

A The minimum value of $y = \frac{1}{3}\sin\theta$ is -3.

B The curve of $y = \sin 3\theta$ oscillates between -3 and 3.

C $y = \tan(\theta - 60°)$ is a periodic function, with a period of 360°.

D The minimum value of $y = \sin 2\theta$ is -1.

2 Which of the following is the curve of $y = \tan x + 2$ for $0° \leqslant x \leqslant 180°$?

A
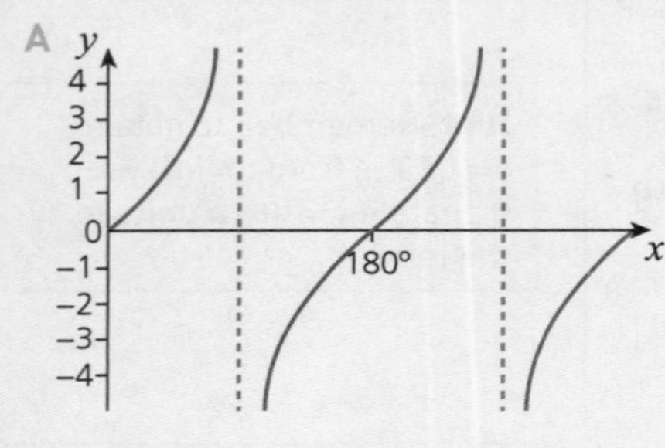

B
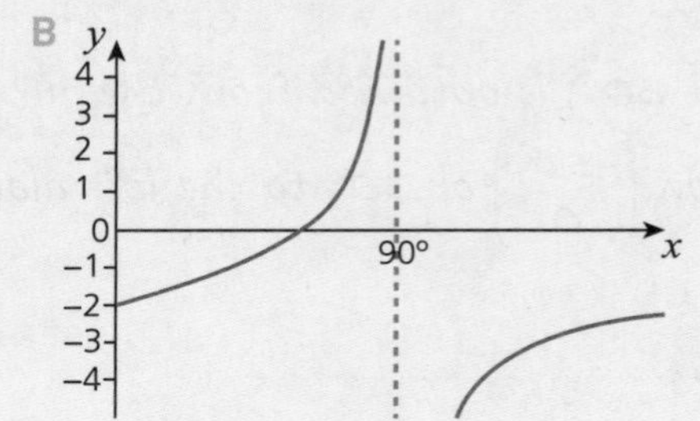

C
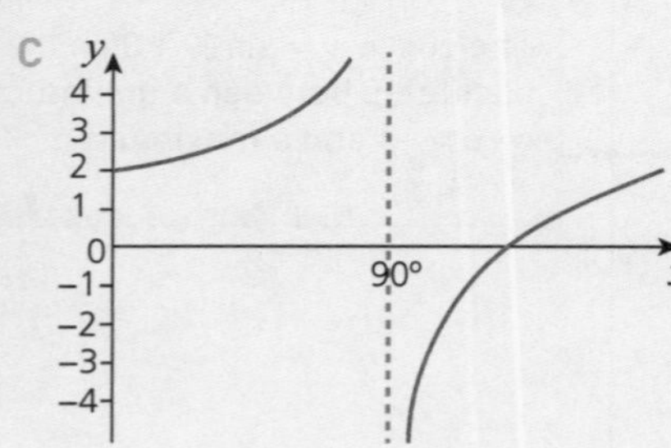

D
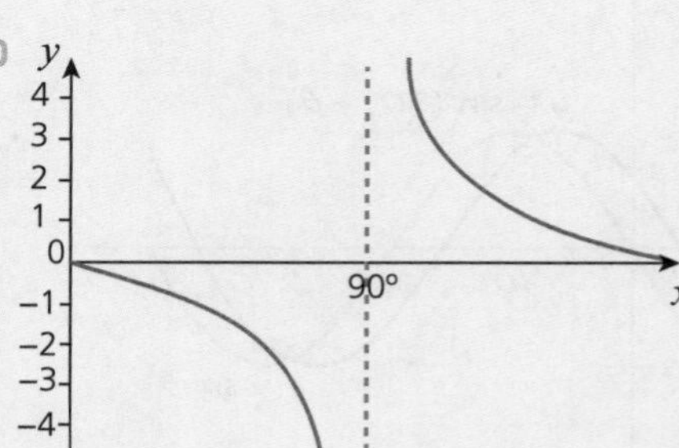

3 Which of the following is the curve of $3y = \cos x$ for $-180° \leqslant x \leqslant 180°$?

A

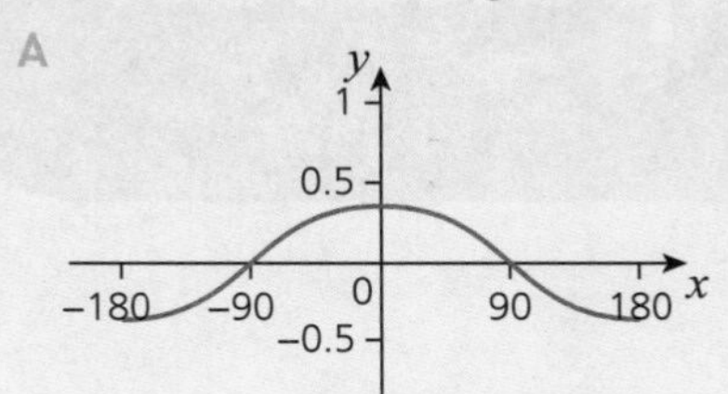

B

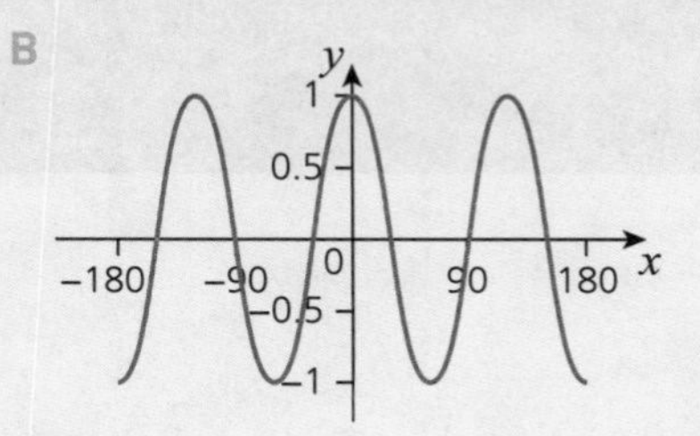

C

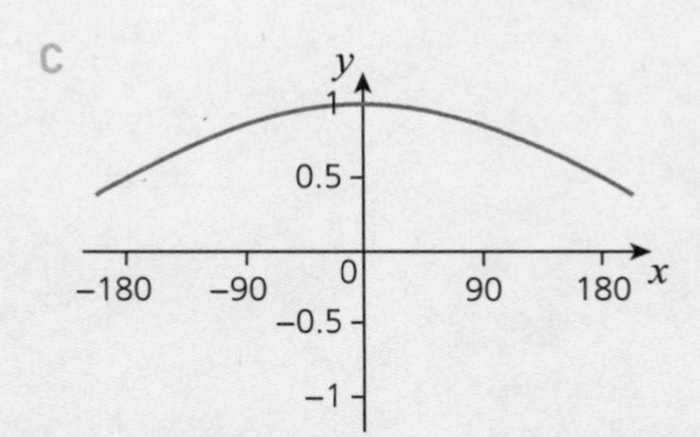

D

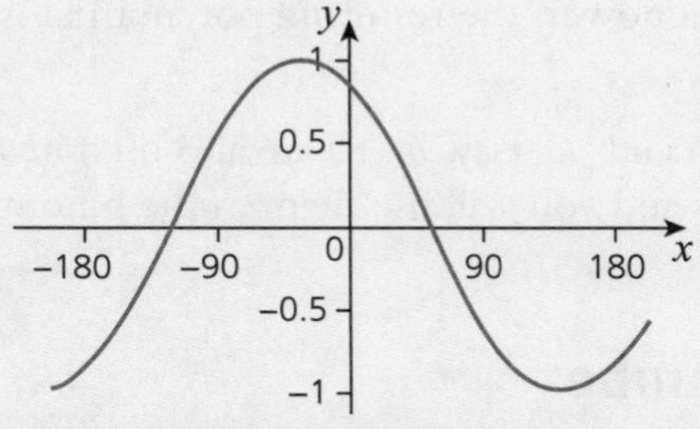

4 Four of the statements below are false and one is true. Which one is true?

A The graph of $y = \sin(x - 90°)$ is the same as the graph of $y = \cos x$.

B The graph below is the same as the graph of $y = \sin(x + 60°)$.

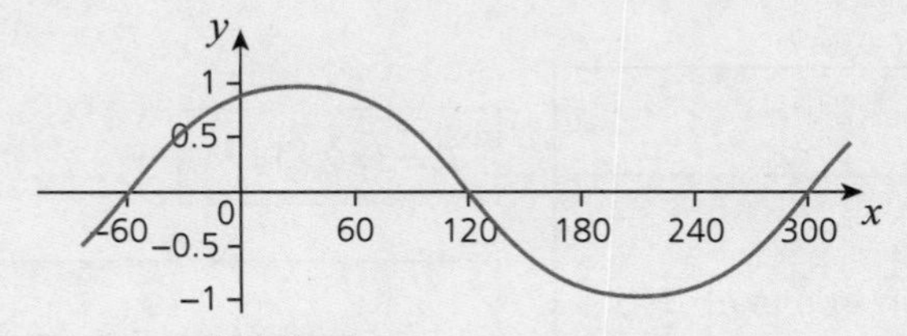

C The graph of $y = -\cos x$ is the same as the graph of $y = \cos x$.

D The graph of $y = \sin(x + 180°)$ is the same as the graph of $y = \sin x$.

E The graph of $y = \sin(x - 45°)$ is the same as the graph of $y = \cos(x + 45°)$.

Full worked solutions online

CHECKED ANSWERS

Exam-style question

i Sketch the curves
a $y = \cos 2x$, **b** $y = \cos(x + 90°)$
on the same axes for $-180° \leqslant x \leqslant 180°$.

ii How many roots does the equation $\cos 2x = \cos(x + 90°)$ have in the interval $-180° \leqslant x \leqslant 180°$?

iii Show that the curves $y = \cos 2x$ and $y = \cos(x + 90°)$ intersect when $y = 0.5$.
Hence solve $\cos 2x = \cos(x + 90°)$ for $-180° \leqslant x \leqslant 180°$.

iv Write down the maximum value of $\cos(x + 90°) - \cos 2x$.

Short answers on page 154

Full worked solutions online

CHECKED ANSWERS

Chapter 9 Binomial expansion

About this topic

A binomial expression is one with two parts such as $(1 + x)$. When a binomial expression is raised to a power, the resulting polynomial is called a binomial expansion.

The notations $n!$ (say, 'n factorial') and ${}_nC_r$ (say, 'n c r') are also used in statistics to work out probabilities and you will use them in the binomial distribution.

Before you start, remember

- how to expand brackets.

Binomial expansions and selections

REVISED

Key facts

1. $n! = n(n-1)(n-2)\ldots\ldots \times 3 \times 2 \times 1$.
   $0! = 1$ and $1! = 1$
2. The number of ways of arranging n unlike objects in line is $n!$
3. The number of possible selections (combinations) of r objects from n unlike objects is ${}_nC_r = \dfrac{n!}{r!(n-r)!}$

   You should use ${}_nC_r$ when the order in which the objects are selected doesn't matter.

   - You may see ${}_nC_r$ written as ${}^nC_r$ or $\begin{pmatrix} n \\ r \end{pmatrix}$
   - ${}_nC_0 = {}_nC_n = 1$
4. A **binomial expression** is an expression with two terms like $(x + 2)$ or $(3x - y)$.
   Here are some binomial expressions raised to a power and their expansions:

$$(x+a)^2 = 1x^2 + 2ax + 1a^2$$

$$(x+a)^3 = 1x^3 + 3ax^2 + 3a^2x + 1a^3$$

$$(x+a)^4 = 1x^4 + 4ax^3 + 6a^2x^2 + 4a^3x + 1a^4$$

   The coefficients in the expansions above are called **binomial coefficients**.
   Polynomials produced by expanding binomials in this way are called **binomial expansions**.
5. You can use **Pascal's triangle** to find **binomial coefficients**.
   Each number in Pascal's triangle is found by adding the two numbers above it.

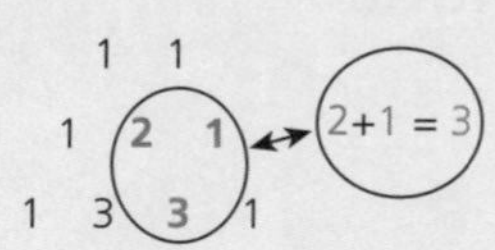

$3! = 3 \times 2 \times 1 = 6$

${}_{10}C_3 = \dfrac{10!}{3! \times (10-3)!} = \dfrac{10!}{3! \times 7!} = 120$

The number of ways of choosing 3 students from a group of 10 to go on a school trip is ${}_{10}C_3 = 120$.
There is 1 way to choose none of the students so ${}_{10}C_0 = 1$ and 1 way to choose all of them so ${}_{10}C_{10} = 1$.

Hint: The powers of the terms in x reduce by one each term and the powers of a increase by one each term. The sum of the two powers is always the same as the power of the bracket.

Hint: Notice that the entries in each row are symmetrical about the middle number.

6 You can also use the formula ${}_nC_r = \dfrac{n!}{r!(n-r)!}$ to calculate **binomial coefficients.**

7 The **binomial theorem** states that:

$$(ax+by)^n = {}_nC_0(ax)^n + {}_nC_1(ax)^{n-1}(by)^1 + {}_nC_2(ax)^{n-2}(by)^2$$
$${}_nC_3(ax)^{n-3}(by)^3 + \dots\dots\dots + {}_nC_n(by)^n$$

Hint: Remember the binomial coefficients are symmetrical and so ${}_nC_r = {}_nC_{n-r}$
For example, ${}_{10}C_3 = \dfrac{10!}{3! \times 7!}$ and ${}_{10}C_7 = \dfrac{10!}{7! \times 3!}$ so both equal 120.

Choosing 3 students out of 10 to go on a trip is the same as choosing which 7 out of 10 will stay behind.

Worked examples

1 Using factorials

Three horses take part in a race. Their names are Arthur's Seat, Blessed Dawn and Cinderella.

How many different ways can they finish?

Solution

There are three possibilities for the first place; for each of these, there are two possibilities for the second place and then only one for the third place.

Hence the total number of ways is $3! = 3 \times 2 \times 1 = 6$

Make sure you can use the factorial button on your calculator.

2 Using ${}_nC_r$ for selections

Jason has 8 different books on his e-reader.

He reads 3 at random on holiday. How many possible selections are there?

Solution

The number of selections is found using ${}_nC_r = \dfrac{n!}{r!(n-r)!}$

$${}_8C_3 = \frac{8!}{3! \times (8-3)!} = \frac{8!}{3! \times 5!} = 56$$

You can use the ${}_nC_r$ button on your calculator to work this out directly.

3 Using Pascal's triangle for binomial expansions (1)

Write out the binomial expansion for $(3a+2b)^4$.

Solution

The binomial coefficients for the fourth row of Pascal's triangle are 1, 4, 6, 4, 1

$$(3a+2b)^4$$
$$= 1(3a)^4 + 4(3a)^3(2b)^1 + 6(3a)^2(2b)^2 + 4(3a)^1(2b)^3 + 1(2b)^4$$
$$= 81a^4 + 4 \times 27a^3 \times 2b + 6 \times 9a^2 \times 4b^2 + 4 \times 3a \times 8b^3 + 16b^4$$
$$= 81a^4 + 216a^3b + 216a^2b^2 + 96ab^3 + 16b^4$$

Common mistake: Remember in the expansion of $(a+bx)^n$ there are $n+1$ terms.

4 Using Pascal's triangle for binomial expansions (2)

Write the binomial expansion of $(x-2)^3$.

Solution

The binomial coefficients for the third row of Pascal's triangle are 1, 3, 3, 1

So: $(x-2)^3 = 1x^3 + 3x^2(-2)^1 + 3x^1(-2)^2 + 1(-2)^3$

$= x^3 - 6x^2 + 12x - 8$

Common mistake: Notice that the negative sign in front of the 2 is treated as part of the number: $x - 2 = x + (-2)$. The end result gives you alternating plus and minus signs in the expansion.

Notice the powers of x are going down – this is in **decreasing powers of x.**

5 Using $_nC_r$ notation for binomial expansions

Write in full the expansion of $(1+x)^4$.

Solution

$(1+x)^4$

$= {}_4C_0 1^4x^0 + {}_4C_1 1^3x^1 + {}_4C_2 1^2x^2 + {}_4C_3 1^1x^3 + {}_4C_4 1^0x^4$

$= \frac{4!}{4!0!}1^4x^0 + \frac{4!}{3!1!}1^3x^1 + \frac{4!}{2!2!}1^2x^2 + \frac{4!}{1!3!}1^1x^3 + \frac{4!}{0!4!}1^0x^4$

$= 1 + 4x + 6x^2 + 4x^3 + x^4$

Hint: Remember ${}_nC_0 = {}_nC_n = 1$

Notice the powers of x are going up – this is in **ascending powers of x.**

6 Calculating single terms in a binomial expansion

What is the term in x^4 in the expansion of $(2+3x)^6$?

Solution

The term will be

$${}_6C_4(2)^2(3x)^4 = 15 \times (2)^2 \times (3x)^4$$
$$= 15 \times 4 \times 81x^4$$
$$= 4860x^4$$

Use your calculator to find ${}_6C_4$.

Make sure you use brackets.

7 Using the binomial expansion for approximations

i Write the binomial expansion of $(1-2x)^4$.

ii Use the first three terms in the expansion to calculate an approximate value of 0.98^4. Give your answer to 3 s.f.

Solution

i $(1-2x)^4$

$= {}_4C_0 + {}_4C_1(-2x) + {}_4C_2(-2x)^2 + {}_4C_3(-2x)^3 + {}_4C_4(-2x)^4$

$= 1 + 4\times(-2x) + 6\times4x^2 + 4\times(-8x)^3 + 1\times16^4$

$= 1 - 8x + 24x^2 - 32x^3 + 16x^4$

ii $0.98^4 = (1-0.02)^4 = (1-2\times0.01)^4$

Substituting $x = 0.01$ into the first three terms of the expansion gives

$$0.98^4 = (1-0.02)^4 \approx 1 - 8(0.01) + 24(0.01)^2$$
$$\approx 1 - 0.08 + 0.0024$$
$$\approx 0.9224$$
$$\approx 0.922 \text{ (to 3 s.f.)}$$

Common mistakes: Be careful with your signs!

So when $x = 0.01$, $(1-2x) = 0.98$.

Common mistake: The question says to use only the first three terms which are $1 - 8x + 24x^2$

Test yourself

TESTED

1 Eloise is making a list of the Christmas presents she wants. There are 9 items on her list. Her mother tells her she can only ask for 3 of them.
How many different ways can she choose the final 3?

A 6 B 9 C 84 D 504 E 362 880

2 Simplify $(x-1)^3+(x+1)$.

A $2x^3+6x$ B $2x^3$ C $2x^3+2x$ D $2x^3+6x^2+6x+2$ E $8x^3$

3 Find the coefficient of x^5 in the expansion of $(2-x)^8$.

A $-{}_8C_5$ B $-{}_8C_5\times 2^5$ C ${}_8C_5\times 2$ D $-{}_8C_5\times 2^3$ E $-{}_8C_3\times 2^5$

4 Write out the binomial expansion of $(1-3x)^4$.

A $1+12x+54x^2+108x^3+81x^4$ B $1-12x+54x^2-108x^3+81x^4$ C $1-12x-18x^2-12x^3-3x^4$

D $1-12x+108x^2-684x^3+1944x^4$ E $1-3x+9x^2-27x^3+81x^4$

5 Jo knows the binomial expansion of $(1+6x)^{10}$.
She wants to use it to obtain an approximation for 0.97^{10}.
What should she take as a value for x?

A −0.01 B 0.005 C −0.005 D −0.03 E −0.05

Full worked solutions online

CHECKED ANSWERS

Exam-style question

i a Expand $(1-2x)^{10}$ as far as the term in x^4.
b Use your expansion to find 0.98^{10} correct to three decimal places.

ii Find the coefficient of x^4 in the expansion of $(1+5x)(1-2x)^{10}$.

Short answers on page 154

Full worked solutions online

CHECKED ANSWERS

Review questions (Chapters 5–9)

1 Three points A, B and C have coordinates (–1, 1), (3, 3) and (2, –5) respectively.

i Find the distance AB and BC.

ii Hence show that triangle ABC is right-angled and find the area of triangle ABC.

2 The points A and B have coordinates (–3, 5) and (7, 9).

i Find the equation of the perpendicular bisector of the line segment AB.

ii Find the equation of the circle with radius AB, centre A.

3 i Prove that $\frac{\cos\theta}{1+\sin\theta}+\frac{\cos\theta}{1-\sin\theta}\equiv\frac{2}{\cos\theta}$

ii Hence solve $\frac{\cos\theta}{1+\sin\theta}+\frac{\cos\theta}{1-\sin\theta}=4$ for $0° \leqslant \theta \leqslant 360°$

4 Sketch the graph of:

i $y=\frac{1}{x}$ **ii** $y=\frac{1}{x}+2$ **iii** $y=\frac{1}{x+2}$

State clearly the coordinates of any intersections with the axes and the equations of any asymptotes.

5 The first three terms in the expansion of $\left(3-\frac{x}{2}\right)$ are $81+bx+cx^2$.

Find the value of each of the constants n, b and c.

6 You are given $f(x)=2x^3-x^2-25x-12$.

i Show that $(x-4)$ is a factor of $f(x)$.

Hence factorise $f(x)$ fully.

The curve $y=f(x)$ is translated by the vector $\begin{pmatrix}2\\0\end{pmatrix}$ to give the curve $y=g(x)$.

ii Solve $g(x)=0$.

Short answers on page 155

Full worked solutions online

CHECKED ANSWERS

SECTION 3

Target your revision (Chapters 10–13)

1 **Differentiate functions involving powers of x**
Differentiate:
i $y = 3x^2 - 2x + 4$ ii $y = \frac{2}{x^2} + \sqrt{x}$
(see pages 89 and 92)

2 **Find the gradient of a curve at a point**
i Find the gradient of the curve $y = 5x - \frac{8}{\sqrt{x}}$ at the point where $x = 4$.
ii Given $f(x) = \frac{x^2}{\sqrt{x}}$, find $f'(9)$.
(see page 92)

3 **Find the equation of a tangent and a normal to a curve**
i Find the equation of the tangent to the curve $y = 3x^2 - 6x$ at the point (3, 9).
ii Find the equation of the normal to the curve $y = 2\sqrt{x}$ at the point where $x = 4$.
(see page 95)

4 **Identify stationary points**
Find the coordinates of the stationary points of the curve $y = -x^3 + 4x^2 - 4$. and identify their nature.
(see page 98)

5 **Identify where a function is increasing or decreasing**
Find the values of x for which $f(x) = 2x^3 + 3x^2 - 36x$ is an increasing function.
(see page 98)

6 **Sketch the graph of the gradient function**
The diagram shows the graph of $y = f(x)$.
Sketch the gradient function, $y = f'(x)$.

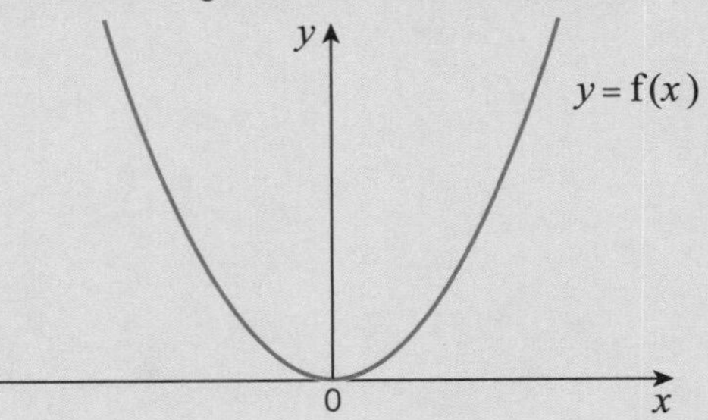

(see page 103)

7 **Find the second derivative**
Given $f(x) = \frac{x^2 + 2\sqrt{x}}{x}$, find $f''(9)$.
(see page 103)

8 **Differentiate from first principles**
i Expand $(x + h)^2$.
ii Given $f(x) = 5x^2$, find an expression for $f(x + h) - f(x)$.
iii Differentiate $y = 5x^2$ from first principles. You must show all your working.
(see page 108)

9 **Find indefinite integrals**
Find:
i $\int (4x^3 - 2x + 3)\,dx$ ii $\int \left(\sqrt{x} - \frac{3}{x^2}\right) dx$
(see pages 112 and 119)

10 **Evaluate definite integrals**
Find:
i $\int_{-2}^{3} (x^2 + 2x - 1)\,dx$ ii $\int_{1}^{4} \left(\frac{12}{x^2} + 6\sqrt{x}\right) dx$
(see pages 115 and 119)

11 **Find the area under a curve**
The diagram shows the graph of $y = x^3 - 2x^2 - 5x + 6$.

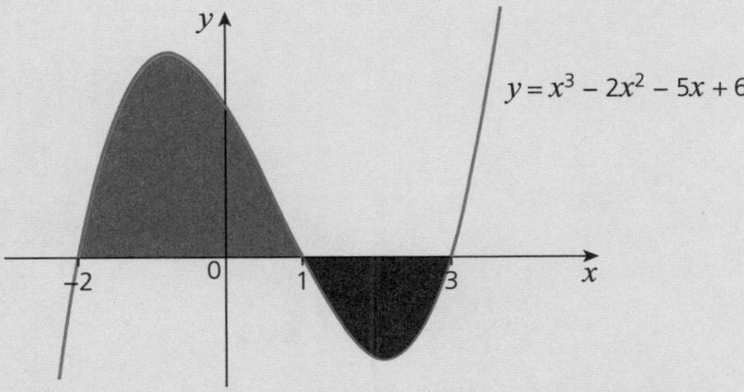

i Find the area of the green shaded region.
ii Find the area of the blue shaded region.
iii Find the total area of the shaded regions.
iv Evaluate $\int_{-2}^{3} (x^3 - 2x^2 - 5x + 6)\,dx$
(see page 115)

12 **Find the magnitude of a vector**
You are given that $\mathbf{q} = \begin{pmatrix} -2 \\ 3 \end{pmatrix}$
The vector **p** has a magnitude of $2\sqrt{13}$ in the same direction as **q**. Find **p**.
(see page 124)

13 Solve problems involving vectors

Three points A, B and C have position vectors $\begin{pmatrix}2\\2\end{pmatrix}$, $\begin{pmatrix}6\\4\end{pmatrix}$ and $\begin{pmatrix}8\\0\end{pmatrix}$ respectively.

i Find $\overrightarrow{AB}$

ii Find $|\overrightarrow{AC}|$

iii Hence show that triangle ABC is right-angled and find the area of triangle ABC.

(see page 124)

14 Sketch the graphs of logarithms and exponentials

Sketch the graph of:

i $y = 1 + \ln x$ ii $y = 2 + e^{-x}$

Label clearly any asymptotes and the coordinates of any intersections with the axes.

(see pages 128 and 132)

15 Simplify expression involving logs

Express $\log x + \log 2 - \log \sqrt{x}$ as a single logarithm.

(see page 128)

16 Solve equations involving logs and exponentials

Solve:

i $2.4^x = 2000$

ii $3\log_{10} x - \log_{10} 20 = \log_{10} 400$

(see pages 128 and 132)

17 Use logs in modelling

The relationship between A and t is modelled by $y = A \times 10^{kt}$, where A and k are constants.

i Show that the graph of $\log y$ against t is a straight line.

The straight line graph obtained when $\log y$ is plotted against t passes through the points (1, 5) and (3, 11). Find

ii the value of A and k

iii the value of y when $t = 2.3$

iv the value of t when $y = 50$.

(see page 137)

Short answers on page 155

Full worked solutions online

CHECKED ANSWERS ☐

Chapter 10 Differentiation

About this topic

You already know how to find the gradient of a straight line. The gradient of a curve is constantly changing, but you can find the gradient of a curve at any point using differentiation.

You can use differentiation to find the equation of a tangent or normal to a point on a curve. You can also use it to find the coordinates of the turning points of a curve and to solve real-life problems.

Before you start, remember

- how to find the equation of a straight line
- laws of indices
- how to sketch graphs of cubic and quadratic functions.

Finding gradients

REVISED

Key facts

1 The gradient of a curve changes as you move along it.
At any point on the curve, the gradient of the curve is the gradient of the tangent to the curve at that point.

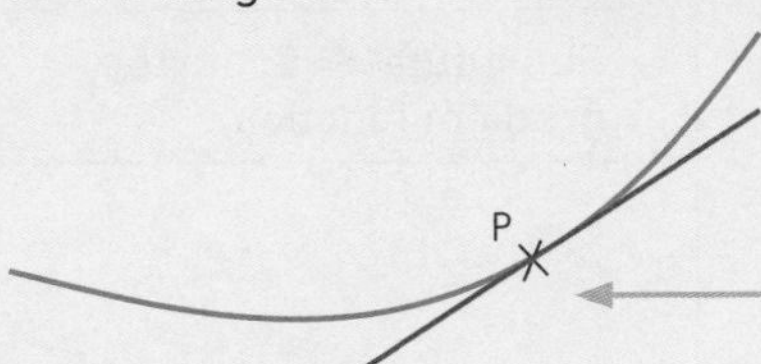

This is the tangent to the curve at point P.

2 A **chord** is a line joining two points on a curve.

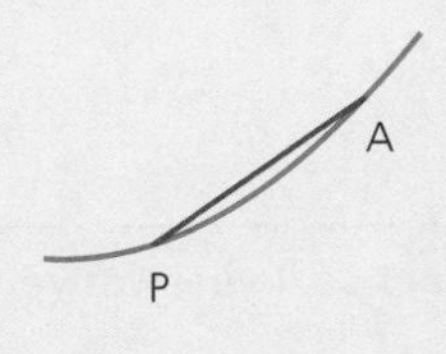

The diagram shows a chord from point P on a curve to another point, A, close to P.
The gradient of the chord PA is an approximation to the gradient of the tangent at P.
You can get a better approximation for the gradient of the tangent at P by using the chord from P to another point B, closer to P than A.

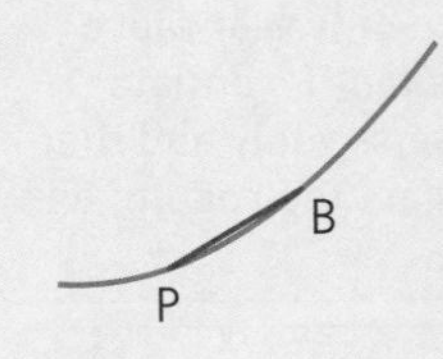

By choosing points closer and closer to point P, you can find chords whose gradients get closer and closer to the gradient of the tangent at P.
The gradient of the tangent at P is the **limit** of the gradient of the chord PA, as point A approaches point P.

Chapter 10 Differentiation

3 The gradient function $\frac{dy}{dx}$ gives the gradient of the curve.

4 The gradient function of $y = kx^n$ is given by $\frac{dy}{dx} = knx^{n-1}$

For example, for the curve $y = x^2$, the gradient function is given by $\frac{dy}{dx} = 2x$

5 The gradient function is sometimes called the **derivative**. As well as telling you the gradient of a curve, the derivative $\frac{dy}{dx}$ tells you the rate of change of y with respect to x. The process of finding a derivative is called **differentiation**.

6 When you differentiate a constant term, k, the result is 0:

$y = k \Rightarrow \frac{dy}{dx} = 0$.

This means that at the point where $x = 1$, the gradient of the curve is 2; at the point where $x = 2$, the gradient of the curve is 4, and so on.

A constant is any number like 2, -0.2, $\frac{1}{2}$ or π.

Hint: This makes sense, since the graph $y = k$ is a horizontal straight line, which has gradient zero at all points.

So when $y = 6$ then $\frac{dy}{dx} = 0$.

Worked examples

1 Finding the gradient of a curve at a point

Find the gradient of the curve $y = x^3$ at the point where $x = 2$.

Solution

$y = x^3 \quad \Rightarrow \frac{dy}{dx} = 3x^2$

When $x = 2$, $\frac{dy}{dx} = 3 \times 2^2 = 3 \times 4 = 12$

The gradient of the curve $y = x^3$ at the point where $x = 2$ is 12.

Hint: $y = kx^n \Rightarrow \frac{dy}{dx} = knx^{n-1}$
You can think of this as 'multiply by the power and reduce the power by 1'.

Substitute $x = 2$ into the gradient function.

2 Differentiating a function

Differentiate the following functions:

i $y = 2x^5$ ii $y = \frac{1}{2}x^{10}$ iii $y = 3$.

Solution

i $y = 2x^5 \quad \Rightarrow \quad \frac{dy}{dx} = 2 \times 5x^4 = 10x^4$

ii $y = \frac{1}{2}x^{10} \quad \Rightarrow \quad \frac{dy}{dx} = \frac{1}{2} \times 10x^9 = 5x^9$

iii $y = 3 \quad \Rightarrow \quad \frac{dy}{dx} = 0$

3 is a constant so its derivative is zero.

3 Differentiating sums and differences of functions

Find the gradient of the function $y = x^4 - 3x^2 + x + 2$ at the point (2, 8).

Solution

$y = x^4 - 3x^2 + x + 2$

$\Rightarrow \frac{dy}{dx} = 4x^3 - (3 \times 2x) + 1 + 0 = 4x^3 - 6x + 1$

When $x = 2$, $\frac{dy}{dx} = (4 \times 2^3) - (6 \times 2) + 1 = 21$.

The gradient of the curve at the point (2, 8) is 21.

Hint: To differentiate the sum or difference of two or more functions, you differentiate each term separately and then add or subtract the results as appropriate.

Remember that the derivative of $1x$ is 1 and the derivative of 2 is zero.

4 Finding the coordinates of a point with a given gradient

Find the coordinates of the point(s) on the curve $y = x^3 - 3x^2 - 8x + 4$ at which the gradient of the curve is 1.

Solution

$$y = x^3 - 3x^2 - 8x + 4 \Rightarrow \frac{dy}{dx} = 3x^2 - 6x - 8$$

At the point for which the gradient is 1:

$$3x^2 - 6x - 8 = 1$$
$$3x^2 - 6x - 9 = 0$$
$$x^2 - 2x - 3 = 0$$
$$(x - 3)(x + 1) = 0$$
$$x = 3 \text{ or } x = -1$$

When $x = 3$, $y = 3^3 - 3 \times 3^2 - 8 \times 3 + 4 = -20$

When $x = -1$, $y = (-1)^3 - 3 \times (-1)^2 - 8 \times (-1) + 4 = 8$

The points at which the gradient of the curve is 1 are (3, −20) and (−1, 8).

Substitute the x coordinates into the equation of the curve to find the y coordinates.

5 Differentiation with other variables

Differentiate the following functions:

i $A = \pi r^2$ ii $f(t) = \frac{2t^3 - t^2}{t}$.

Common mistake: In part ii you need to simplify before you can differentiate.

Solution

i $A = \pi r^2 \Rightarrow \frac{dA}{dr} = 2\pi r$

ii $f(t) = \frac{2t^3 - t^2}{t} = 2t^2 - t$

$f'(t) = 4t - 1$

You need the rate of change of A with respect to r.

Hint: When $y = f(x)$ you can use the notation $f'(x)$ instead of $\frac{dy}{dx}$. In this case you have $f(t)$ so you should write $f'(t)$.

Test yourself

TESTED

1 Differentiate $z = t^4 - 2t^3 - t + 3$.

A $\frac{dz}{dt} = 4t^3 - 6t^2 + 2$ B $\frac{dy}{dt} = 4t^3 - 6t^2$ C $\frac{dy}{dx} = 4t^3 - 6t^2 - 1$

D $\frac{dy}{dt} = 4t^3 - 5t^2 - 1$ E $\frac{dz}{dt} = 4t^3 - 6t^2 - 1$

2 Given $f(x) = 4 - 3x - x^2 + 2x^3$, find $f'(-2)$.

A 31 B 25 C 29 D −23 E 21

3 Find the gradient of the curve $y = 3x^2 - 5x - 1$ at the point (2, 1).

A 5 B 7 C 6 D 1 E −1

4 Find the coordinates of the point on the curve $y = 4 - 3x + x^2$ at which the tangent to the curve has gradient −1.

A (1, 2) B (1, −1) C (−2, 14) D (−2, −1)

5 Find the gradient of the curve $y = x^5(2x + 1)$ at the point at which $x = -1$.

A 10 B −3 C −7 D −17 E 7

Full worked solutions online

CHECKED ANSWERS

Exam-style question

A is the point on the curve $y = 2x^2 - 3x + 1$ with x coordinate 2.

B is the point on the same curve with x coordinate 2.1.

i Calculate the gradient of the chord AB of the curve.

ii Give the x coordinate of a point C on the curve for which the gradient of chord AC is a better approximation to the gradient of the curve at A.

iii Find the gradient of the curve at A by differentiation.

Short answers on page 156

Full worked solutions online

CHECKED ANSWERS

Extending the rule

REVISED

Key fact

The gradient function of $y = kx^n$ is given by $\frac{dy}{dx} = knx^{n-1}$

This is true for all values of n, including **negative and fractional values**.

Hint: If you need to differentiate a function of the form $\frac{k}{x^n}$, write it as kx^{-n} first.

Worked examples

1 Differentiating negative powers of x

Find the gradient function for the curve $y = x^{-3}$.

Solution

$$y = x^{-3} \Rightarrow \frac{dy}{dx} = -3x^{-4} = -\frac{3}{x^4}$$

2 Finding the gradient of a curve at a point

Find the gradient of the curve $y = \frac{3}{x^2}$ at the point where $x = 2$.

Solution

$$y = \frac{3}{x^2} = 3x^{-2} \Rightarrow \frac{dy}{dx} = 3 \times -2x^{-3} = -6x^{-3}$$

When $x = 2$, $\frac{dy}{dx} = -6 \times 2^{-3} = \frac{-6}{2^3} = -\frac{6}{8} = -\frac{3}{4}$

The gradient of the curve at the point where $x = 2$ is $-\frac{3}{4}$.

Hint: When differentiating a negative power of x, you use the standard result in the same way that you do for a positive power of x.

Multiply by the power of −3 and subtract 1 from the power: −3 − 1 = −4

$y = x^{-3}$ is the same as $y = \frac{1}{x^3}$ and the answer, $\frac{dy}{dx} = -3x^{-4}$ is the same as $\frac{dy}{dx} = -\frac{3}{x^4}$.

3 Differentiating sums and differences of functions

Differentiate the function $y = \frac{1}{x} - \frac{2}{x^4} + 3$.

Solution

$$y = \frac{1}{x} - \frac{2}{x^4} + 3 = x^{-1} - 2x^{-4} + 3$$

Rewrite the function using negative indices.

$$\frac{dy}{dx} = -1x^{-2} - 2 \times -4x^{-5} + 0$$

Differentiate each term. Be careful with signs.

$$= -x^{-2} + 8x^{-5}$$

$$= -\frac{1}{x^2} + \frac{8}{x^5}$$

If you want to you can write the final expression using fractions.

4 Differentiating fractional powers of x (1)

Differentiate the function $y = x^{\frac{2}{3}}$.

Hint: When differentiating a fractional power of x, you use the standard result in the same way that you do for an integer power of x.

Solution

$$y = x^{\frac{2}{3}} \quad \Rightarrow \frac{dy}{dx} = \frac{2}{3}x^{-\frac{1}{3}}$$

The value of n is $\frac{2}{3}$, so the value of $n - 1$ is: $\frac{2}{3} - 1 = -\frac{1}{3}$.

5 Differentiating fractional powers of x (2)

Find the gradient of the curve $y = 2\sqrt{x^3}$ at the point at which $x = 4$.

Common mistakes: If you need to differentiate a function involving a root, write it using a fractional index first.

Remember $x^{\frac{1}{n}} = \sqrt[n]{x}$. So, write $\sqrt[3]{x}$ as $x^{\frac{1}{3}}$, and $\frac{1}{\sqrt{x}}$ as $x^{-\frac{1}{2}}$.

Solution

$$y = 2\sqrt{x^3} = 2x^{\frac{3}{2}}$$

$\sqrt{x^3} = (x^3)^{\frac{1}{2}} = x^{\frac{3}{2}}$

$$\frac{dy}{dx} = 2 \times \frac{3}{2}x^{\frac{1}{2}}$$

$$= 3x^{\frac{1}{2}}$$

$$= 3\sqrt{x}$$

If you want to, you can write the final expression using a square root. This may make it easier when you substitute in a value for x.

When $x = 4$, $\frac{dy}{dx} = 3\sqrt{4} = 3 \times 2 = 6$

The gradient of the curve at the point at which $x = 4$ is 6.

6 Differentiating surds

Differentiate the function $y = \sqrt[3]{x} - \frac{4}{\sqrt{x}}$.

Solution

$$y = \sqrt[3]{x} - \frac{4}{\sqrt{x}}$$

$$= x^{\frac{1}{3}} - 4x^{-\frac{1}{2}}$$

Rewrite the expression using fractional indices.

$$\frac{dy}{dx} = \frac{1}{3}x^{-\frac{2}{3}} - 4 \times -\frac{1}{2}x^{-\frac{3}{2}}$$

Differentiate each term. Be careful with signs.

$$= \frac{1}{3}x^{-\frac{2}{3}} + 2x^{-\frac{3}{2}}$$

Hint: You could write the answer using square roots, but it is fine to leave it as it is.

7 Differentiating term by term

Find the derivative of $y = \frac{1}{2x^3} - 3\sqrt{x} + x^2 - 2$.

Solution

$$y = \frac{1}{2x^3} - 3\sqrt{x} + x^2 - 2x$$

$$= \frac{1}{2}x^{-3} - 3x^{\frac{1}{2}} + x^2 - 2x$$

$$\frac{dy}{dx} = \frac{1}{2} \times -3x^{-4} - 3 \times \frac{1}{2}x^{-\frac{1}{2}} + 2x - 2$$

$$= -\frac{3}{2}x^{-4} - \frac{3}{2}x^{-\frac{1}{2}} + 2x - 2$$

$$= -\frac{3}{2x^4} - \frac{3}{2\sqrt{x}} + 2x - 2$$

Hint: When you differentiate you are finding the derivative.

Common mistakes: You are less likely to make a mistake when you differentiate if you rewrite the expression using negative and fractional indices first.

Test yourself

TESTED

1 Differentiate the function $y = \frac{1}{3x^5}$.

A $\frac{dy}{dx} = -\frac{15}{x^6}$ B $\frac{dy}{dx} = -\frac{5}{3x^4}$ C $\frac{dy}{dx} = -\frac{15}{x^4}$ D $\frac{dy}{dx} = -\frac{5}{3x^6}$ E $\frac{dy}{dx} = \frac{1}{15x^4}$

2 Differentiate the function $y = 3\sqrt[3]{x^2}$.

A $\frac{dy}{dx} = 2x^{-\frac{2}{3}}$ B $\frac{dy}{dx} = x^{-\frac{2}{3}}$ C $\frac{dy}{dx} = 2x^{-\frac{1}{3}}$ D $\frac{dy}{dx} = 8x^{\frac{5}{3}}$ E $\frac{dy}{dx} = -18x^{-7}$

3 Find the gradient of the function $y = x - 2 + \frac{3}{x} - \frac{2}{x^2}$ at the point where $x = -2$.

A $-\frac{1}{4}$ B $\frac{5}{4}$ C $\frac{3}{4}$ D $\frac{7}{2}$ E -4

4 Find the gradient of the function $y = 5\sqrt{x} - \frac{4}{\sqrt{x}}$ at the point where $x = 4$.

A $\frac{19}{12}$ B 1 C $\frac{3}{2}$ D $-\frac{59}{4}$ E $\frac{9}{4}$

5 Find the coordinates of the point(s) at which the graph $y = x - \frac{1}{x}$ has gradient 5.

A $\left(\frac{1}{2}, -\frac{3}{2}\right)$ B $\left(2, \frac{3}{2}\right)$ and $\left(-2, -\frac{3}{2}\right)$ C $\left(2, \frac{3}{2}\right)$ D $\left(-\frac{1}{2}, \frac{3}{2}\right)$ E $\left(\frac{1}{2}, -\frac{3}{2}\right)$ and $\left(-\frac{1}{2}, \frac{3}{2}\right)$

Full worked solutions online

CHECKED ANSWERS

Exam-style question

i Differentiate $y = x - \frac{3}{x^4} + \frac{2}{x^5}$.

ii Differentiate $y = 3x^2 - 2\sqrt{x^5} + \frac{1}{\sqrt[3]{x}}$.

Short answers on page 156

Full worked solutions online

CHECKED ANSWERS

Tangents and normals

REVISED

Chapter 10 Differentiation

Key facts

1. You can find the **gradient, m_1, of a tangent** to a curve at a given point using differentiation.
2. The **equation of the tangent** to a curve at the point (x_1, y_1) is given by $y - y_1 = m_1(x - x_1)$.
3. The **normal** to a curve at a given point is the straight line which is perpendicular to the tangent at that point.

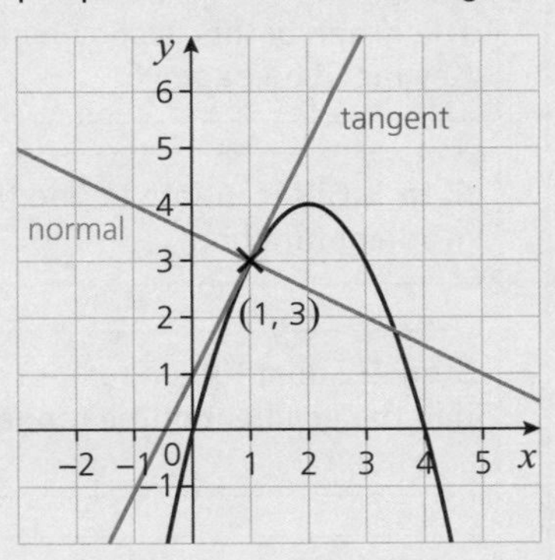

The diagram shows the curve $y = 4x - x^2$ and the tangent and normal at the point (1, 3).

4. The **gradient, m_2, of a normal** to a curve at a given point can be found by first finding the gradient m_1 of the tangent, and then using the relationship $m_1m_2 = -1$.
5. The **equation of the normal** to a curve at the point (x_1, y_1) is given by $y - y_1 = m_2(x - x_1)$.

Worked examples

1 Finding the equation of a tangent to a curve

The diagram shows the curve $y = x^2 - x + 1$ and the tangent at the point (2, 3).

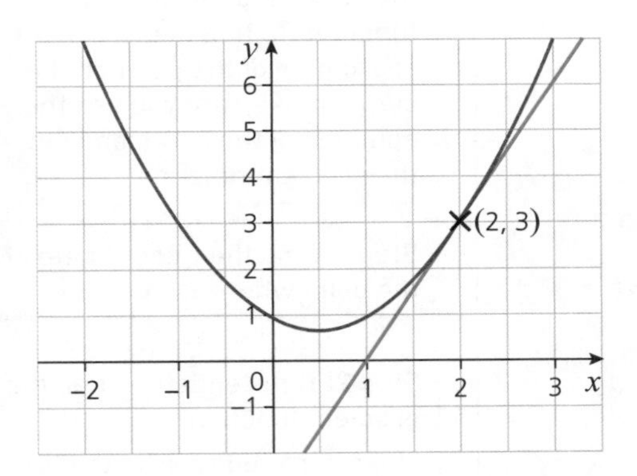

Find the equation of this tangent.

Solution

$y = x^2 - x + 1 \quad \Rightarrow \frac{dy}{dx} = 2x - 1$

When $x = 2$, $\frac{dy}{dx} = 2 \times 2 - 1 = 4 - 1 = 3$

The tangent has gradient 3 and passes through the point (2, 3)

so the equation of the tangent is $y - 3 = 3(x - 2)$

$y - 3 = 3x - 6$

$y = 3x - 3$

Step 1: Differentiate to find the gradient function.

Step 2: Substitute in for x to find the gradient of the tangent at a point.

Step 3: Use $y - y_1 = m_1(x - x_1)$ to find the equation of the tangent.

The graph of this is shown in Key fact 3 on page 95.

2 Finding the equation of a normal to a curve

Find the equation of the normal to the curve $y = 4x - x^2$ at the point (1, 3).

Solution

$y = 4x - x^2 \quad \Rightarrow \frac{dy}{dx} = 4 - 2x$

When $x = 1$, the gradient m_1 of the tangent is $m_1 = 4 - 2 \times 1 = 2$

The gradient m_2 of the normal is $m_2 = -\frac{1}{m_1} = -\frac{1}{2}$

The normal has gradient $-\frac{1}{2}$ and passes through the point (1, 3)

$x_1 = 1, y_1 = 3$

so the equation of the normal is $y - y_1 = m_2(x - x_1)$

$y - 3 = -\frac{1}{2}(x - 1)$

$2y - 6 = -(x - 1)$

$2y - 6 = -x + 1$

$2y + x = 7$

Step 1: Differentiate to find the gradient function.

Step 2: Substitute in for x to find the gradient of the tangent at a point.

Step 3: Find the gradient of the normal.

The gradient of the normal is the negative reciprocal of the gradient of the tangent.

Step 4: Use $y - y_1 = m_2(x - x_1)$ to find the equation of the normal.

Common mistake: Sometimes you may only be given x_1, the x coordinate of the point at which you need to find a tangent or normal. If this is the case, you will need to find y_1 first.

Don't mix up the equation of the curve and the gradient function. To find the y coordinate of the point on the curve, make sure you use the equation of the curve and not the gradient function!

3 Finding the equation of a tangent and a normal to a curve

Find the equations of the tangent and normal to the curve $y = \frac{8}{x^2} - \sqrt{x}$ at the point where $x = 4$.

Solution

When $x = 4$, $y = \frac{8}{4^2} - \sqrt{4} = \frac{8}{16} - 2 = -\frac{3}{2}$, so the point is $\left(4, -\frac{3}{2}\right)$

$y = \frac{8}{x^2} - \sqrt{x} \Rightarrow y = 8x^{-2} - x^{\frac{1}{2}} \quad \Rightarrow \frac{dy}{dx} = -16x^{-3} - \frac{1}{2}x^{-\frac{1}{2}}$

$= -\frac{16}{x^3} - \frac{1}{2\sqrt{x}}$

When $x = 4$, the gradient m_1 of the tangent is

$m_1 = -\frac{16}{4^3} - \frac{1}{2\sqrt{4}} = -\frac{1}{4} - \frac{1}{4} = -\frac{1}{2}$

Step 1: Find the y coordinate of the point where $x = 4$.

Step 2: Differentiate to find the gradient function

Make sure you rewrite the function using fractional and negative powers first.

Hint: It is easier to substitute in for x when you write the answer in this form.

Step 3: Substitute $x = 4$ to find the gradient of the tangent.

So the tangent has gradient $-\frac{1}{2}$ and passes through the point $\left(4, -\frac{3}{2}\right)$

So $x_1 = 4$ and $y_1 = -\frac{3}{2}$

The equation of the tangent is $y - y_1 = m_1(x - x_1)$

Step 4: Find the equation of the tangent.

$$y - \left(-\tfrac{3}{2}\right) = -\frac{1}{2}(x - 4)$$

$$2y + 3 = -x + 4$$

$$2y + x = 1$$

Hint: It is easier to deal with the fraction by multiplying through by 2.

The gradient m_2 of the normal is $m_2 = -\dfrac{1}{m_1} = -\dfrac{1}{-\frac{1}{2}} = 2$

Step 5: Find the gradient of the normal.

The normal has gradient 2 and passes through the point $\left(4, -\frac{3}{2}\right)$

Step 6: Find the equation of the normal.

So the equation of the normal is $y - y_1 = m_2(x - x_1)$

$$y - \left(-\tfrac{3}{2}\right) = 2(x - 4)$$

$$y + \frac{3}{2} = 2x - 8$$

$$2y + 3 = 4x - 16$$

Multiply through by 2.

$$2y - 4x + 19 = 0$$

Test yourself

TESTED

1 Find the gradient of the normal to the curve $y = \dfrac{4}{\sqrt{x}}$ at the point where $x = 4$.

A $-\dfrac{1}{4}$ B -4 C 4 D 3 E -1

2 Find the equation of the tangent to the curve $y = x^3 - 3x^2 + x + 4$ at the point where $x = 1$.

A $y + 2x = 5$ B $y + 2x = 1$ C $y = -2x$ D $y + 2x = 7$ E $y + 2x = 3$

3 Find the equation of the normal to the curve $y = x^2 + 7x + 6$ at the point where $x = -2$.

A $y = 3x + 2$ B $3y + 10 = x$ C $3y + x + 38 = 0$ D $y + 6 = 3x$ E $3y + x + 14 = 0$

4 Find the equation of the tangent to the curve $y = \sqrt{x}$ at the point where $x = 4$.

A $4y = x - 8$ B $y + x = 6$ C $4y = x + 8$ D $4y = x - 2$ E $4y = x + 4$

5 Find the equation of the normal to the curve $y = \dfrac{1}{x}$ at the point where $x = 2$.

A $y + x = 4$ B $2y = 8x - 15$ C $2y + 8x = 1$ D $4y = x$ E $2y = 8x - 17$

Full worked solutions online

CHECKED ANSWERS

Exam-style question

The diagram shows the cubic curve with equation $y = 2x^3 - 3x + 1$.

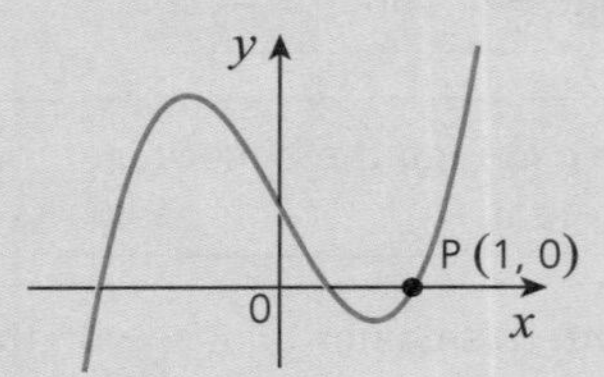

i Show that the tangent to the curve at the point P (1, 0) has gradient 3.
ii Find the coordinates of the other point, Q, on the curve at which the tangent has gradient 3.
iii Find the equation of the normal to the curve at Q.

Short answers on page 156

Full worked solutions online

CHECKED ANSWERS

Increasing and decreasing functions, and turning points

REVISED

Key facts

1 A function is increasing if its gradient function is positive.
If the gradient function is positive everywhere, the function is called an **increasing function**.

So y increases as x increases.

2 A function is decreasing if its gradient function is negative.
If the gradient function is negative everywhere, the function is called a **decreasing function**.

So y decreases as x increases.

3 Many functions are increasing functions for some values of x, and decreasing functions for other values of x.

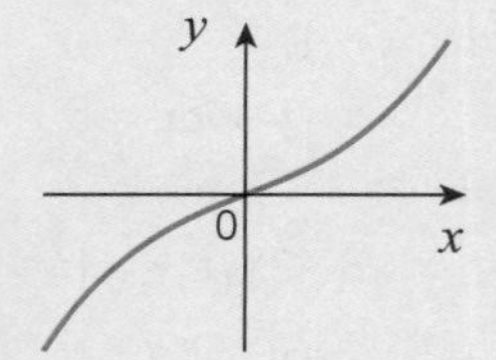

This is an increasing function

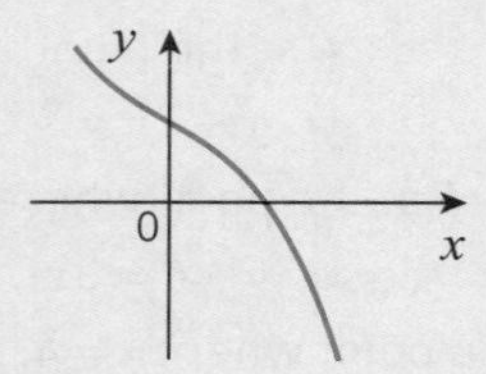

This is a decreasing function

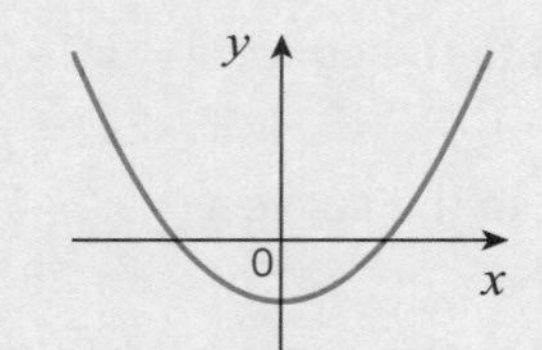

This function is decreasing for negative values of x, and increasing for positive values of x

4 **Stationary points** on a curve are points at which the gradient of the curve is zero.
This means that the tangent to the curve is horizontal and $\frac{dy}{dx} = 0$ at that point.
A stationary point can be:

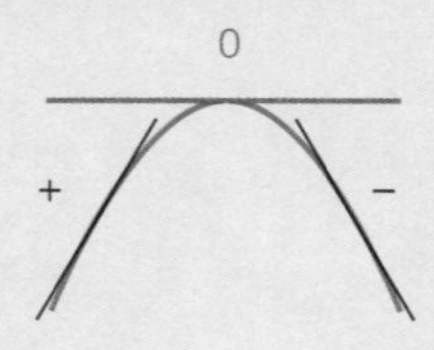

a local maximum, at which the gradient changes from positive to negative

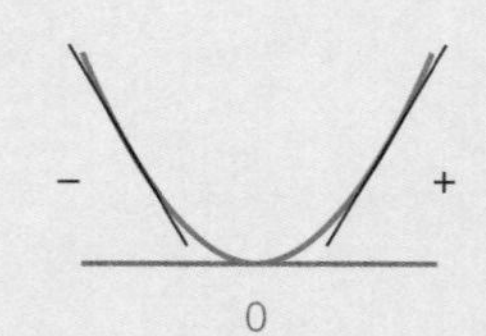

a local minimum, at which the gradient changes from negative to positive

At a maximum or a minimum, the curve turns, so these are also called **turning points.**

5 You can find the coordinates of a stationary (or turning) point by
- finding $\frac{dy}{dx}$
- and then finding the value(s) of x for which $\frac{dy}{dx} = 0$
- and then substituting the value(s) of x into the equation of the curve to find the y coordinates.

$y = \ldots$

6 The nature of a stationary (or turning) point can be found by considering the sign of the $\frac{dy}{dx}$ on either side of the point.

The gradient function.

At a local **minimum point** $\smile$
$\frac{dy}{dx}$ is **negative** to the left of the point and **positive** to the right of the point

At a local **maximum point** $\frown$
$\frac{dy}{dx}$ is **positive** to the left of the point and **negative** to the right of the point

Worked examples

1 Finding where a function is increasing

Find the range of values of x for which the function $y = x^3 - 3x + 1$ is an increasing function of x.

Solution

$y = x^3 - 3x + 1 \quad \Rightarrow \frac{dy}{dx} = 3x^2 - 3$

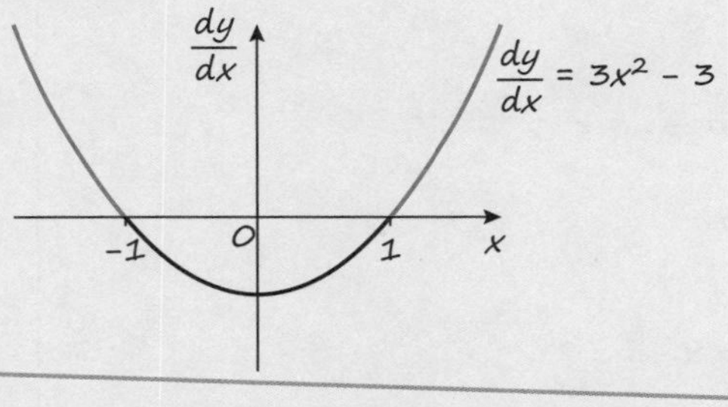

$\frac{dy}{dx}$

$\frac{dy}{dx} = 3x^2 - 3$

-1 O 1 x

The function is increasing if

$\frac{dy}{dx} > 0$

$3x^2 - 3 > 0$

$x^2 - 1 > 0$

$(x - 1)(x + 1) > 0$

This is the difference of two squares $x^2 - a^2 = (x - a)(x + a)$

A sketch of the graph of $\frac{dy}{dx} = 3x^2 - 3$ shows that the solution to this inequality is $x < -1$ or $x > 1$.

So the function is increasing when $x < -1$ or $x > 1$.

2 Finding stationary points (1)

i Find the coordinates of the stationary points on the curve $y = x^3 - 3x^2 - 9x + 10$.

ii Find the nature of the stationary points.

iii Sketch the curve.

Hint: It is a good idea to set your work out in a table, as shown in this example.

Solution

i $y = x^3 - 3x^2 - 9x + 10 \quad \Rightarrow \frac{dy}{dx} = 3x^2 - 6x - 9$

Step 1: Differentiate

At stationary points, $\frac{dy}{dx} = 0$, so $3x^2 - 6x - 9 = 0$

$$x^2 - 2x - 3 = 0$$
$$(x - 3)(x + 1) = 0$$
$$x = 3 \text{ or } x = -1$$

Step 2: Set $\frac{dy}{dx} = 0$ and solve.

When $x = 3$, $\quad y = 3^3 - 3 \times 3^2 - 9 \times 3 + 10 = -17$

When $x = -1$, $y = (-1)^3 - 3(-1)^2 - 9(-1) + 10 = 15$

The stationary points are (3, −17) and (−1, 15).

Step 3: Substitute the x-values into the equation of the curve to find the y coordinate.

ii At the point where $x = -2$:

$$\frac{dy}{dx} = 3(-2)^2 - 6(-2) - 9 = 15 > 0$$

At the point where $x = 0$:

$$\frac{dy}{dx} = 3 \times 0 - 6 \times 0 - 9 = -9 < 0$$

At the point where $x = 4$:

$$\frac{dy}{dx} = 3 \times 4^2 - 6 \times 4 - 9 = 15 > 0$$

Hint: Examine the sign of $\frac{dy}{dx}$ on either side of the turning point.

Note: You only need to determine whether $\frac{dy}{dx}$ is positive or negative, the exact value of $\frac{dy}{dx}$ doesn't matter.

	$x < -1$	$x = -1$	$-1 < x < 3$	$x = 3$	$x > 3$
Sign of $\frac{dy}{dx}$	+ve /	0 —	−ve \	0 —	+ve /
Stationary point		Local maximum (0; + ∩ −)		Local minimum (0; − ∪ +)	

Common mistakes: Be careful if the stationary points are close together. For example, if there are stationary points at $x = \frac{1}{2}$ and $x = 1$, then you can't use $x = 0$ to look at the gradient on the left of the point $x = 1$. You need to use a point between $x = \frac{1}{2}$ and $x = 1$, such as $x = \frac{3}{4}$.

So (−1, 15) is a local maximum point, and (3, −17) is a local minimum point.

iii

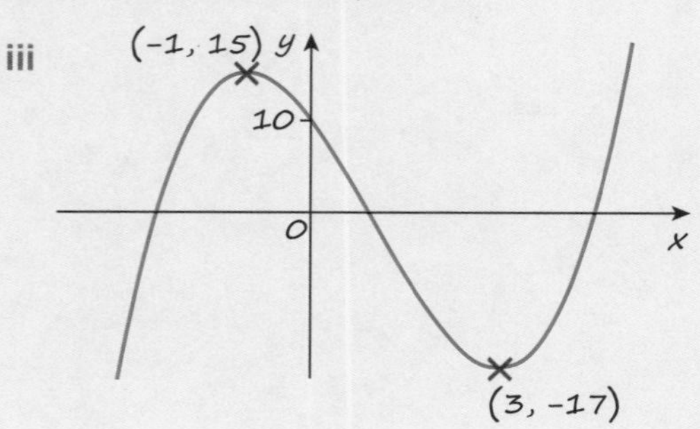

You can find where the curve intersects the y-axis by substituting $x = 0$ into the equation of the curve. In this case, when $x = 0$ then $y = 10$.

In this example, you weren't required to find the coordinates of the points where the curve intersects the x-axis. However, in the exam you may be asked to label all the points where the curve intersects the axes.

3 Finding stationary points (2)

Find the coordinates of the turning points on the curve $y = 2x - \frac{1}{x^2}$, $x \neq 0$ and determine their nature.

You can't divide by 0 so the function isn't defined at this point.

Solution

$$y = 2x - \frac{1}{x^2}$$

$$\Rightarrow y = 2x - x^{-2}$$

$$\Rightarrow \frac{dy}{dx} = 2 - (-2)x^{-3}$$

$$= 2 + \frac{2}{x^3}$$

Common mistake: Take care with your signs!

Step 1: Differentiate

At a turning point, $\frac{dy}{dx} = 0$, so

$$2 + \frac{2}{x^3} = 0$$

$$\Rightarrow \frac{2}{x^3} = -2$$

$$\Rightarrow x^3 = -1$$

$$\Rightarrow x = -1$$

Step 2: Set $\frac{dy}{dx} = 0$ and solve.

When $x = -1$, $y = 2 \times (-1) - \frac{1}{(-1)^2} = -2 - 1 = -3$

Step 3: Substitute $x = -1$ into the equation of the curve to find the y coordinate.

The turning point is at (−1, −3).

At the point where $x = -2$:

$$\Rightarrow \frac{dy}{dx} = 2 + \frac{2}{(-2)^3} = 2 - \frac{1}{4} = 1.75 > 0$$

At the point where $x = -\frac{1}{2}$:

$$\Rightarrow \frac{dy}{dx} = 2 + \frac{2}{(\frac{1}{-2})^3} = 2 - 16 = -14 < 0$$

Step 4: Examine the sign of $\frac{dy}{dx}$ on either side of the turning point.

	$x < -1$	$x = -1$	$x > -1$
Sign of $\frac{dy}{dx}$	+ve	0	−ve
Stationary point		Local maximum	

So (−1, −3) is a local maximum point.

Test yourself

TESTED

1 Look at the three functions below:

$f(x) = x^3$

$g(x) = x^3 + x$

$h(x) = x^2 + x$

Which of these functions are increasing functions of x for all values of x?

A f only B f and g C g only D f and h E h only

2 Find the range of values of x for which the function $y = x^3 + 2x^2 + x$ is a decreasing function of x.

A $x < -1$ and $x > -\frac{1}{3}$ B $x \leqslant -1$ and $x \geqslant -\frac{1}{3}$ C $x < -2$ D $-1 \leqslant x \leqslant -\frac{1}{3}$ E $-1 < x < -\frac{1}{3}$

3 Find the coordinates of the stationary point(s) on the curve $y = x^3 - 3x^2 - 9x + 11$.

A (1, 0) and (−3, −16) B (−1, 0) and (3, 0) C (−1, 16) and (3, −16)

D (1, −12) and (−3, 36) E (1, 0) only

4 Find the coordinates of the stationary point(s) on the curve $y = 3x^4 + 2x^3 + 1$.

A $\left(-\frac{1}{2}, \frac{15}{16}\right)$ only B $(0, 1)$ and $\left(-\frac{1}{2}, \frac{9}{16}\right)$ C $\left(-\frac{1}{2}, \frac{9}{16}\right)$ only

D $(0, 1)$ and $\left(-\frac{1}{2}, \frac{15}{16}\right)$ E $(0, 1)$ and $\left(\frac{1}{2}, \frac{23}{16}\right)$

5 Find the x coordinates of the turning point(s) on the curve $y = x + \frac{1}{x^3}, x \neq 0$ and identify the nature of each turning point.

A There is a local maximum at $x = -\sqrt[4]{3}$ and a local minimum at $x = \sqrt[4]{3}$.

B There is a local minimum at $x = \sqrt[4]{3}$ only.

C There is a local minimum at $x = -\sqrt[4]{3}$ and a local maximum at $x = \sqrt[4]{3}$.

D There is a local maximum at $x = -\sqrt{3}$ and a local minimum at $x = \sqrt{3}$.

E There is a local minimum at $x = \sqrt{3}$ only.

Full worked solutions online

CHECKED ANSWERS

Exam-style question

A curve has equation $y = x^3 - 3x^2 - 9x + 2$.

i Find $\frac{dy}{dx}$.

ii Find the range of values of x for which y is an increasing function of x.

iii Find the coordinates of the stationary points on the curve $y = x^3 - 3x^2 - 9x + 2$ and determine their nature.

Short answers on page 156

Full worked solutions online

CHECKED ANSWERS

Higher derivatives and the graph of $\frac{dy}{dx}$

REVISED

Key facts

1 You can use the graph of $y = f(x)$ to sketch the graph of the gradient function.
- Look for any **stationary points**, where $\frac{dy}{dx} = 0$
- Look for regions where the function is **increasing**, where $\frac{dy}{dx} > 0$
- Look for regions where the function is **decreasing**, where $\frac{dy}{dx} < 0$

2 The second derivative is found by differentiating the gradient function $\frac{dy}{dx}$ or $f'(x)$.

It is written as $\frac{d^2y}{dx^2}$ or $f''(x)$

The second derivative tells you the rate of change of the gradient function.

3 You can use the sign of $\frac{d^2y}{dx^2}$ at a stationary point to determine the nature of that stationary point:
- If $\frac{d^2y}{dx^2} > 0$, it is a local minimum.
- If $\frac{d^2y}{dx^2} < 0$, it is a local maximum.
- if $\frac{d^2y}{dx^2} = 0$, you need to look at the gradient on either side to find out the nature of the stationary point.

The graph of $\frac{dy}{dx}$ will **cross** the x-axis at the points where $y = f(x)$ has a **stationary point**.

The graph of $\frac{dy}{dx}$ will **lie above** the x-axis at the points where $y = f(x)$ is **increasing**.

The graph of $\frac{dy}{dx}$ will **lie below** the x-axis at the points where $y = f(x)$ is **decreasing**.

You will learn more about this special case in A level Mathematics.

Worked examples

1 Sketching the graph of the gradient function

The diagram shows the graph of $y = f(x)$.

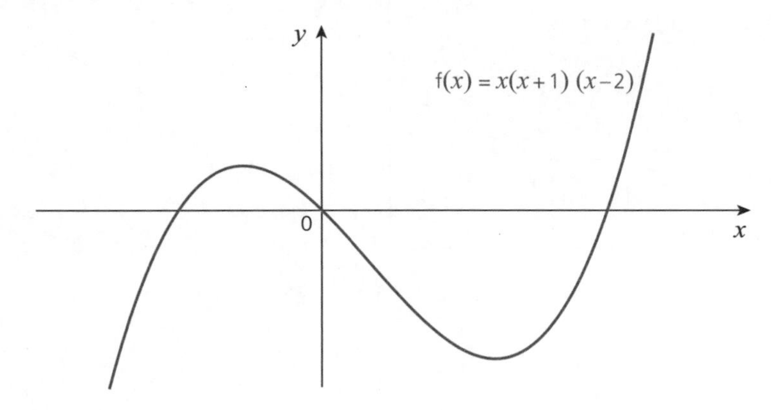

Sketch the graph of $y = f'(x)$.

Solution

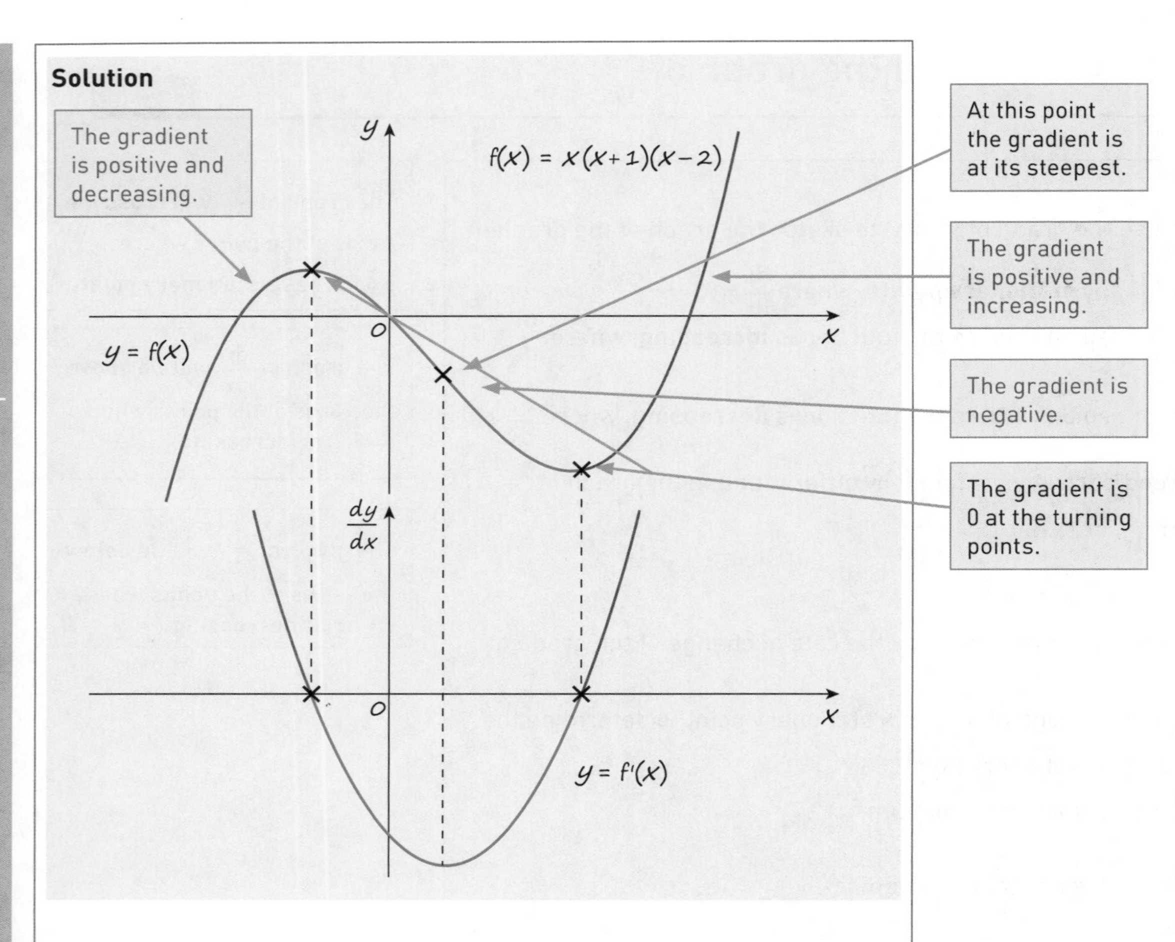

2 Finding the second derivative

Find the second derivative of the function $y = x^4 - 3x^2 - 2x + \frac{1}{x}$.

Solution

$$y = x^4 - 3x^2 - 2x + x^{-1}$$

Differentiate to give the gradient function:

$$\frac{dy}{dx} = 4x^3 - 6x - 2 - x^{-2}$$

Differentiate again to give the second derivative:

$$\frac{d^2y}{dx^2} = 12x^2 - 6 - 0 + 2x^{-3}$$

$$= 12x^2 - 6 + \frac{2}{x^3}$$

3 Finding the value of the second derivative at a point

Given $f(x) = x^3 - x^2 - 2\sqrt{x}$.

Find the values of f'(4) and f"(4).

Solution

$$f(x) = x^3 - x^2 - 2x^{\frac{1}{2}}$$

Rewrite using fractional powers of x.

$$\Rightarrow f'(x) = 3x^2 - 2x - x^{-\frac{1}{2}}$$

$$\Rightarrow f'(x) = 3x^2 - 2x - \frac{1}{\sqrt{x}}$$

$$\Rightarrow f''(x) = 6x - 2 + \frac{1}{2}x^{-\frac{3}{2}}$$

$$\Rightarrow f''(x) = 6x - 2 + \frac{1}{2\sqrt{x^3}}$$

Hint: Rewrite using square roots to make it easier to substitute in values of x.

When $x = 4$, $f'(x) = 3 \times 4^2 - 2 \times 4 - \frac{1}{\sqrt{4}} = 48 - 8 - \frac{1}{2} = 39\frac{1}{2}$

When $x = 4$, $f''(x) = 6 \times 4 - 2 + \frac{1}{2\sqrt{4^3}} = 24 - 2 + \frac{1}{16} = 22\frac{1}{16}$

4 Using the second derivative to determine the nature of stationary points

The curve $y = 2x^5 + 5x^4 - 1$ has stationary points at (–2, 15) and (0, –1).

Determine the nature of these stationary points.

Solution

$$y = 2x^5 + 5x^4 - 1 \Rightarrow \frac{dy}{dx} = 10x^4 + 20x^3$$

$$\Rightarrow \frac{d^2y}{dx^2} = 40x^3 + 60x$$

When $x = -2$, $\frac{d^2y}{dx^2} = 40(-2)^3 + 60(-2)^2 = -320 + 240 = -80$

Since the second derivative is negative, (-2, 15) is a local maximum point.

When $x = 0$, $\frac{d^2y}{dx^2} = 40 \times 0^3 + 60 \times 0^2 = 0$

Since the second derivative is zero, test the gradient function on either side.

When $x = -1$, $\frac{dy}{dx} = 10(-1)^4 + 20(-1)^3 = 10 - 20 = -10$

When $x = 1$, $\frac{dy}{dx} = 10 \times 1^4 + 20 \times 1^3 = 10 + 20 = 30$

The gradient function is going from negative to positive, so (0, -1) is a local minimum point.

Test yourself

TESTED

1 The diagram shows the graph of $y = f(x)$.

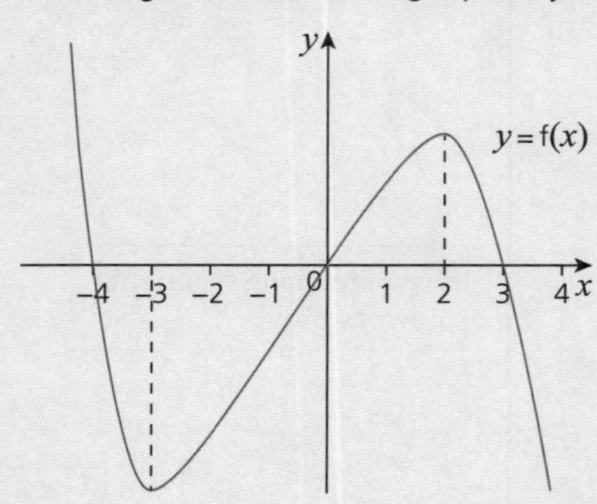

Which of the following shows the graph of $y = f'(x)$?

A

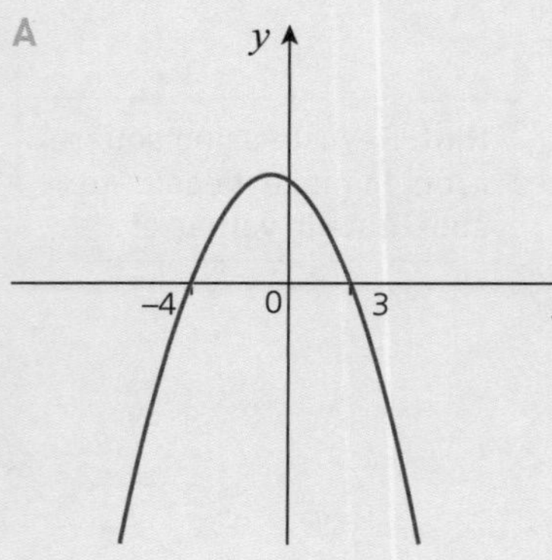

B

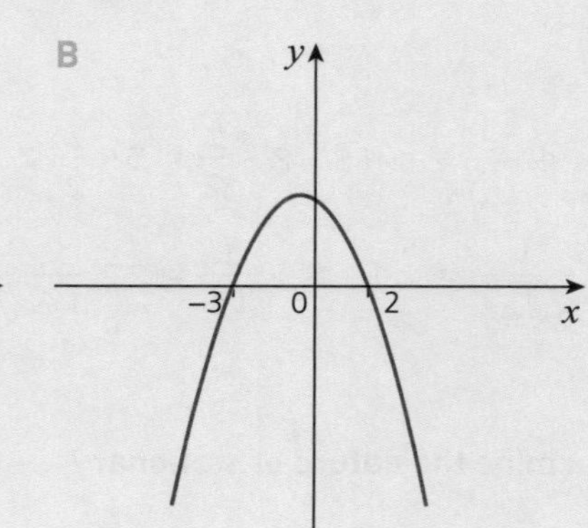

C

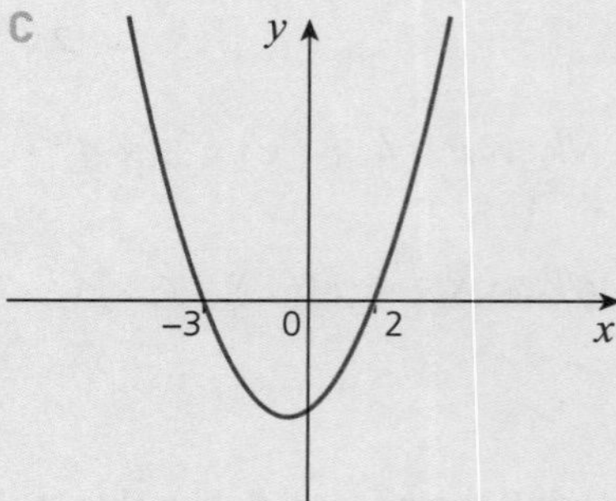

D

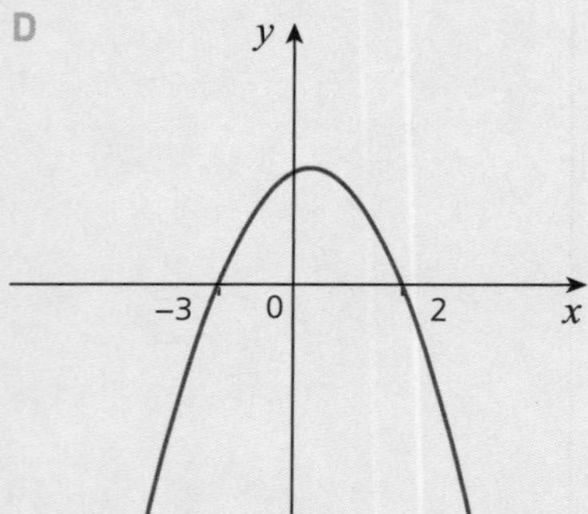

E

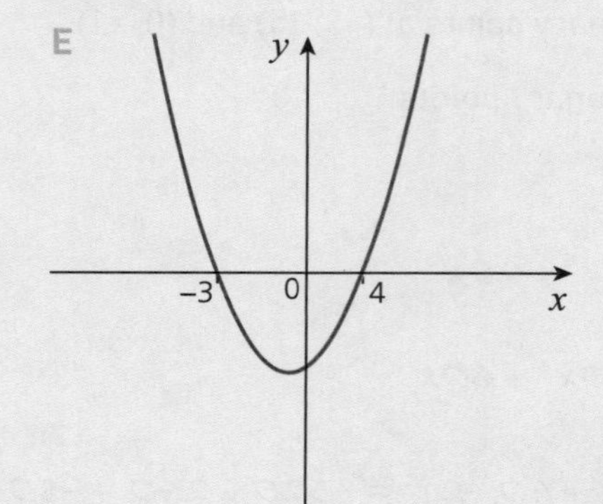

2 Find the value of the second derivative of the curve $y=x^4-3x^2-x+1$ at the point where $x=-2$.

A -21 B 42 C -54 D 10 E -18

3 Find the value of the second derivative of the curve $y=\frac{1}{3x}$ at the point where $x=3$.

A $\frac{2}{81}$ B $-\frac{2}{81}$ C $\frac{2}{9}$ D $-\frac{1}{27}$ E $-\frac{2}{9}$

4 Find the y coordinate of the point on the curve $y=x^3+6x^2+5x-3$ at which the second derivative is zero.

A -2 B -45 C 3 D -7 E 9

5 Find the x coordinate of the stationary point of the curve $y=x-\sqrt{x}+2$.
Use the second derivative to identify its nature.

A $x=\frac{1}{4}$; minimum B $x=\frac{1}{4}$; maximum C $x=\sqrt{2}$; minimum

D $x=\frac{1}{\sqrt{2}}$; minimum E $x=\frac{1}{\sqrt{8}}$; maximum

Full worked solutions online

CHECKED ANSWERS

Exam-style question

i The diagram shows the graph of $y=\mathrm{f}(x)$.

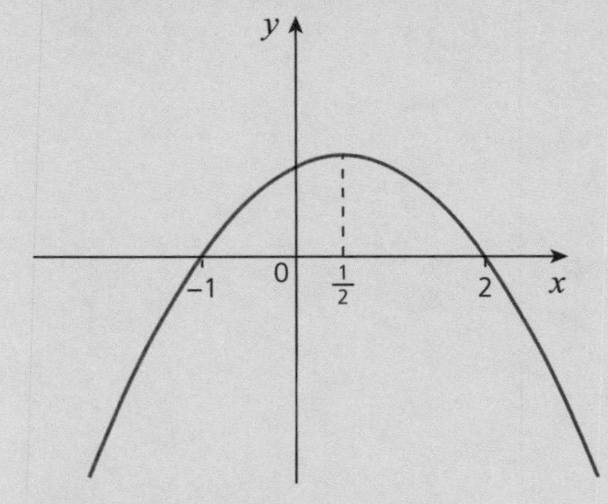

Sketch the graph of $y=\mathrm{f}'(x)$.
Label the coordinates of the point(s) where the graph crosses the x-axis.

ii The curve $y=2x\sqrt{x}-\sqrt{x}$ has one turning point at P.
Find the x coordinate of P and use the second derivative to identify the nature of the turning point at P.

Short answers on page 157

Full worked solutions online

CHECKED ANSWERS

Applications and differentiation from first principles

REVISED

Key facts

1 You can use differentiation to solve practical problems in which the maximum or minimum value of a quantity is needed.

Step 1: Write an equation for the quantity, say A, to be optimised

Step 2: Rewrite the equation so that A is in terms of just one other variable, say w.

Step 3: Differentiate to find $\frac{dA}{dw}$

Step 4: Solve $\frac{dA}{dw}=0$ to find the value of w that gives a maximum or minimum value of A.

Step 5: Substitute this value of w into the equation found in step 2 to give the corresponding value of A.

See Example 1.

2 You need to be able to differentiate small integer powers of x from first principles.

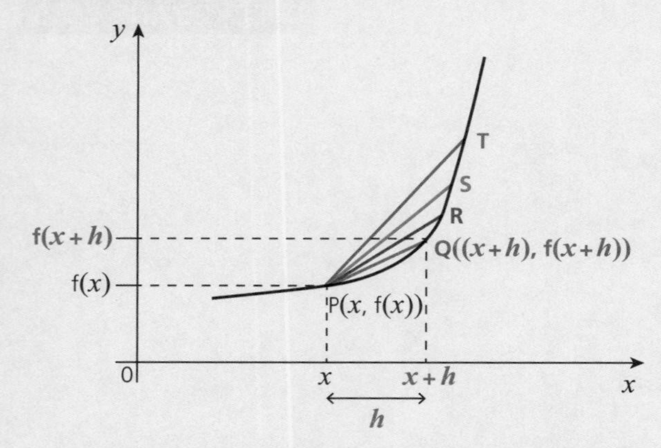

This diagram shows that the gradient of a series of chords from P converges to the gradient of the tangent at P.

You can see the gradient of **chord PS** is a better approximation for the gradient at P than **chord PT** and the gradient of **chord PR** is better still.

The closer a point is to P the closer the gradient of the chord is to the gradient of the tangent at P. So, the gradient of **chord PQ** where **point Q**, is only a short distance, h, away from P is very close to the gradient of the tangent at P.

The gradient of **chord PQ** is $\frac{f(x+h)-f(x)}{h}$.

Gradient is 'Rise over run'.

Say 'As h tends to 0'.

As $h \to 0$ then the gradient of chord PQ $\to$ the gradient of the tangent at P.

So you can write $f'(x)=\lim_{h\to 0}\frac{f(x+h)-f(x)}{h}$.

When h is very close to 0, you can say the gradient function at P is equal to the gradient of chord PQ.

Worked examples

1 Using differentiation to solve real-life problems

A cardboard box is a cuboid with a height that is twice its width.

The volume of the cuboid is 1000 cm^3.

Find the width of the cuboid so that the surface area is a minimum.

Find the minimum surface area and demonstrate that this is a minimum.

Solution

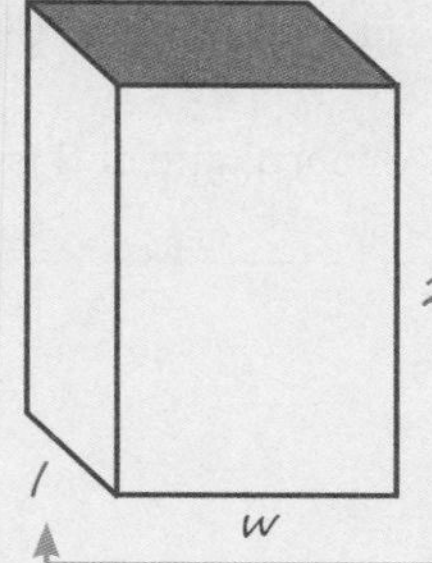

The height is twice the width.

The question doesn't tell you anything about the length so just call this side l.

$$\text{Volume} = wl \times 2w = 1000$$

$$\Rightarrow 2w^2 l = 1000 \quad ①$$

Step 1: Write down an equation for the surface area.

$$\text{Surface area, } A = 2 \times 2w^2 + 2 \times wl + 2 \times 2wl$$

$$= 4w^2 + 6wl \quad ②$$

Step 2: You need to re-write this so it is just in terms of w before you differentiate.

Make l the subject of ①...

$$\text{From ①: } l = \frac{1000}{2w^2} = \frac{500}{w^2}$$

... then substitute into ② so you have A in terms of w.

$$\text{From ②: } A = 4w^2 + 6w \times \frac{500}{w^2}$$

$$\Rightarrow A = 4w^2 + \frac{3000}{w}$$

$$\Rightarrow A = 4w^2 + 3000w^{-1}$$

Step 3: Now you can differentiate.

$$\text{Differentiating gives } \frac{dA}{dw} = 8w - 3000w^{-2}$$

$$= 8w - \frac{3000}{w^2}$$

$$\text{At a minimum, } \frac{dA}{dw} = 0 \Rightarrow 8w - \frac{3000}{w^2} = 0$$

$$\Rightarrow 8w = \frac{3000}{w^2}$$

$$\Rightarrow 8w^3 = 3000$$

$$\Rightarrow w^3 = 375$$

$$\Rightarrow w = 7.21 \text{ cm to 3 s.f.}$$

Step 4: Substitute $w = 7.21...$ into $A = 4w^2 + \frac{3000}{w}$ to find the minimum surface area.

When $w = 7.21...$ then $A = 4 \times 7.21...^2 + \frac{3000}{7.21...} = 624$ cm^2 to 3 s.f.

At a minimum, $\frac{d^2A}{dw^2} > 0$.

To show this is a minimum differentiate again:

$$\frac{d^2A}{dw^2} = 8 + 6000w^{-3}$$

When $w = 7.21$ then $\frac{d^2A}{dw^2} = 8 + \frac{6000}{7.21...^3} = 24 > 0$ and so this is a minimum.

Common mistakes: Don't round values until you reach your final answer.

Make sure you answer the question fully!

2 Differentiating from first principles

Differentiate $f(x) = 3x^2$ from first principles.

Solution

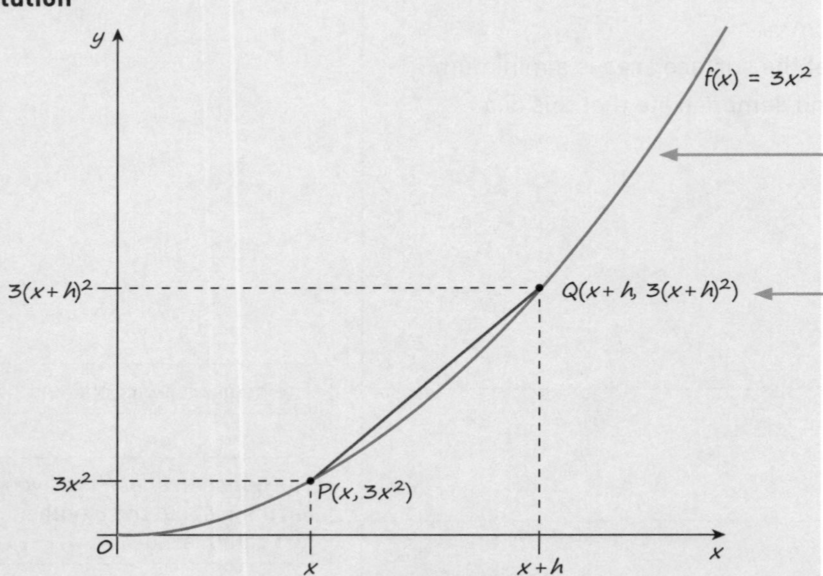

Start by drawing a sketch showing the point P on the curve $y = f(x)$ and the point Q a distance h away from P.

The curve is $y = 3x^2$ and so the y coordinate of P is $3x^2$...

...and the y coordinate of Q is $3(x + h)^2$.

The gradient of chord PQ $= \dfrac{3(x+h)^2 - 3x^2}{h}$ — Gradient is 'rise over run'.

$= \dfrac{3(x^2 + 2xh + h^2) - 3x^2}{h}$ — Expand the brackets.

$= \dfrac{3x^2 + 6xh + 3h^2 - 3x^2}{h}$ — The $3x^2$ terms cancel.

$= \dfrac{6xh + 3h^2}{h}$ — Cancel h.

$= 6x + 3h$

When $h \to 0$, the gradient of chord PQ $\to$ the gradient of the tangent at P — This is the derivative of $f(x) = 3x^2$.

When $h \to 0$, $6x + 3h \to 6x$ — When h is very close to 0, $3h$ is also very close to 0.

So when $f(x) = 3x^2$, $f'(x) = 6x$

Test yourself

TESTED ☐

Questions 1 and 2 are to do with the following situation.

A rectangular sheet of sides 24 cm and 15 cm has four equal squares of sides x cm cut from the corners.

The sides are then turned up to make an open rectangular box.

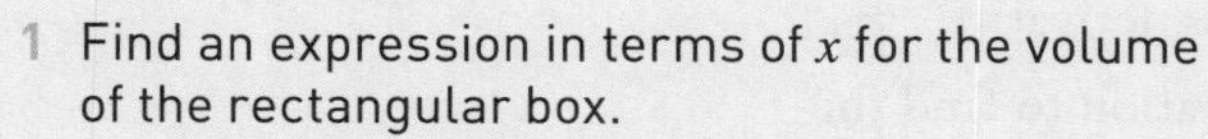

1 Find an expression in terms of x for the volume of the rectangular box.

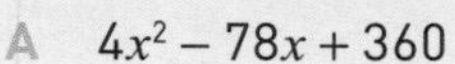

A $4x^2 - 78x + 360$ B $360x$

C $4x^3 - 78x^2 + 360x$ D $2x^3 - 39x^2 + 180x$

2 Find the value of x so that the volume of the box is a maximum.

A 3 B 10 C 9.75 D 2.55

Questions 3–5 are to do with the following graph.

P and Q are two points on the curve $y = x^3$.

P is the point (x, x^3) and the x coordinate of Q is $(x + h)$.

3 Find the expansion of $(x + h)^3$.

A $x^3 + h^3$ B $x^3 + 6x^2h + h^3$

C $x^3 + 3x^2h + 3h^2x + h^3$ D $x^3 + 3xh + 3hx + h^3$

4 Find an expression for the gradient of the chord PQ.

A $\dfrac{x^3 + 3x^2h + 3hx^2 + h^3}{h}$ B $\dfrac{3x^2h + 3h^2x + h^3}{x + h}$

C $3x^2$ D $3x^2 + 3hx + h^2$

5 Write down the limit of the gradient of the chord as $h \to 0$.

A 0 B $3x^2 + 3x$ C x^3 D $3x^2$

Full worked solutions online

CHECKED ANSWERS ☐

Exam-style question

A cylindrical waste paper bin is made from a thin sheet of metal. The bin has no lid.

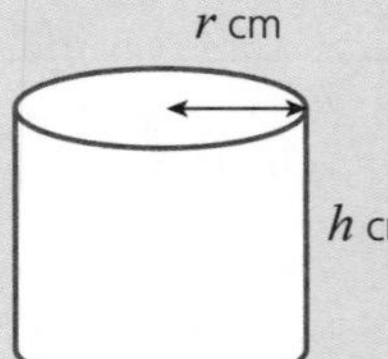

Hint: 15 litres = 15 000 cm³.

The bin needs to hold 15 litres and has a height h cm and radius r cm

Find the minimum surface area of metal required to make the bin and prove that this value is the minimum.

Short answers on page 157

Full worked solutions online

CHECKED ANSWERS ☐

Chapter 11 Integration

About this topic

Integration is the opposite process to differentiation. You can use integration to find the equation of a curve if you know its derivative, $\frac{dy}{dx}$, and a point that it passes through. You can also use integration to find the area bounded by some curves and the x-axis.

Before you start, remember

- differentiation
- how to sketch the graph of a quadratic or cubic equation
- the laws of indices.

Integration as the reverse of differentiation

REVISED

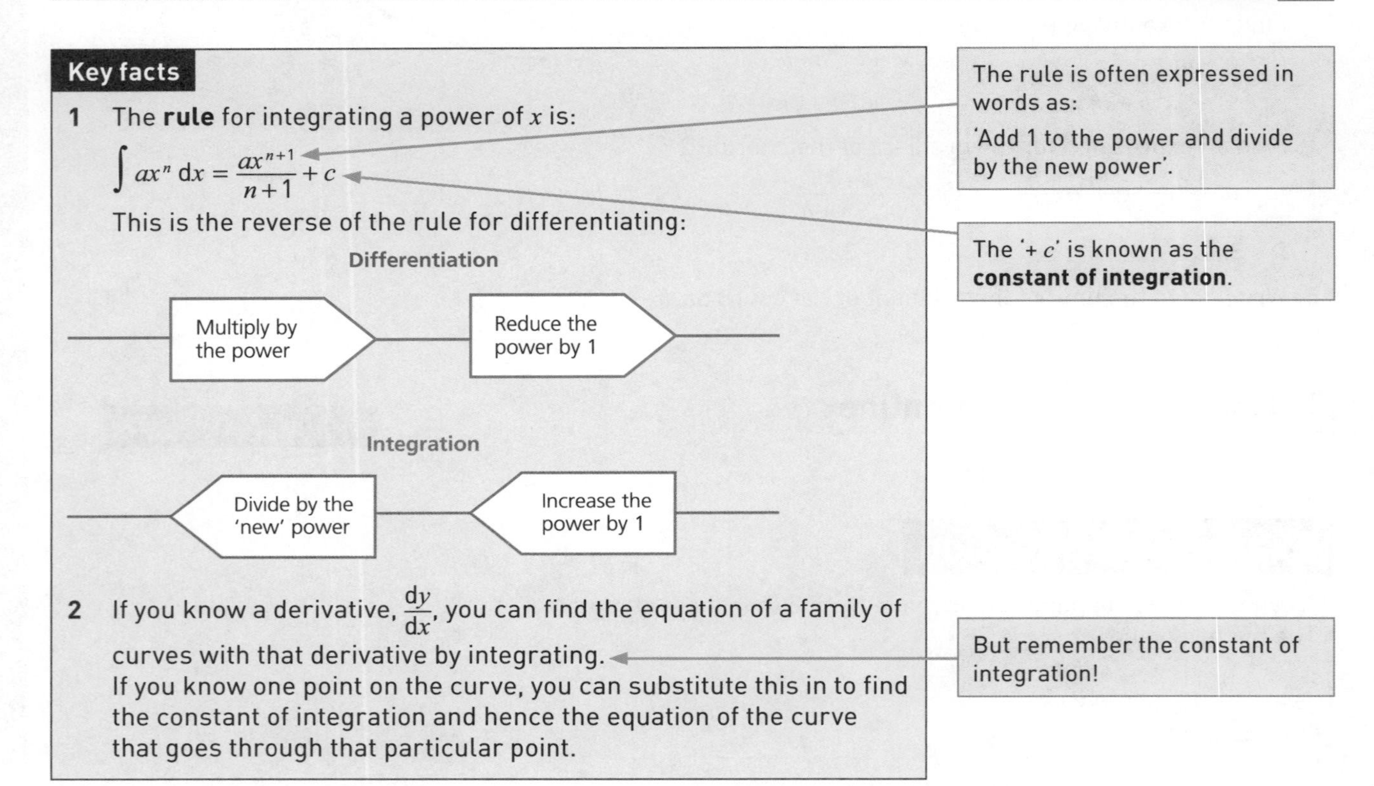

Working examples

1 Integrating powers of x

Find:

i $\int 12x^2\,dx$

ii $\int 5x\,dx$

iii $\int (3x^2+10x-1)\,dx$

iv $\int 3\,dx$.

Solution

i $\int 12x^2\,dx = \frac{12x^3}{3} + c$

$= 4x^3 + c$

ii $\int 5x\,dx = \frac{5x^2}{2} + c$

iii $\int (3x^2+10x-1)\,dx = \frac{3x^3}{3} + \frac{10x^2}{2} - x + c$

$= x^3 + 5x^2 - x + c$

iv $\int 3\,dx = 3x + c$

Add 1 to the power and divide by the new power.

Don't forget the '$+c$'.

Integrate each term separately and then add them (the same as you'd do for differentiation).

Common mistakes: Do not get confused integrating a constant term, or number. You can think of this in two ways:

- $3x$ differentiates to 3, so 3 integrates to $3x$ (because integration is the opposite of differentiation)
- 3 can be written as $3x^0$ which would integrate to $\frac{3x^1}{1}$ or $3x$.

2 Finding the equation of a curve given its gradient function, $\frac{dy}{dx}$

The gradient function of a curve is $\frac{dy}{dx} = 2x$. The curve passes through (2, 7). Find the equation of the curve.

Solution

Integrating $\frac{dy}{dx} = 2x$ gives $y = \frac{2x^2}{2} + c$

Simplifying: $y = x^2 + c$

Substitute the point (2, 7) into the equation:

$7 = 2^2 + c$

$\Rightarrow 7 = 4 + c$

$\Rightarrow c = 3$

So the equation of the curve is $y = x^2 + 3$

This is the **general equation** of the curve.

This is the **particular solution** that passes through the point (2, 7).

3 Finding and using the equation of a curve given its gradient function, $\frac{dy}{dx}$

A curve passes through (−1, 8) and its gradient is given by

$\frac{dy}{dx} = 3x^2 + 6x - 9.$

Find the equation of the curve and sketch the graph of y against x showing the coordinates of the intercept with the y-axis and any stationary points.

Solution

Integrating gives: $y = \frac{3x^3}{3} + \frac{6x^2}{2} - 9x + c$

Simplifying gives: $y = x^3 + 3x^2 - 9x + c$.

Substitute the point (−1, 8) into the equation of the curve:

$8 = (-1)^3 + 3 \times (-1)^2 - 9 \times (-1) + c$

$\Rightarrow 8 = -1 + 3 + 9 + c$

$\Rightarrow c = -3$

So the equation of the curve is $y = x^3 + 3x^2 - 9x - 3$

So the curve cuts the y-axis at (0, −3).

The graph intercepts the y-axis at $y = -3$.

The stationary points on the curve are where $\frac{dy}{dx} = 0$,

$3x^2 + 6x - 9 = 0$

$\Rightarrow 3(x^2 + 2x - 3) = 0$

$\Rightarrow 3(x + 3)(x - 1) = 0$

$\Rightarrow x = -3$ or $x = 1$

Substituting these into the equation of the curve gives:

Common mistake: Don't forget to find the y coordinates of the turning points.

$y = (-3)^3 + 3 \times (-3)^2 - 9 \times (-3) - 3$

$= 24$

$y = 1^3 + 3 \times 1^2 - 9 \times 1 - 3$

$= -8$

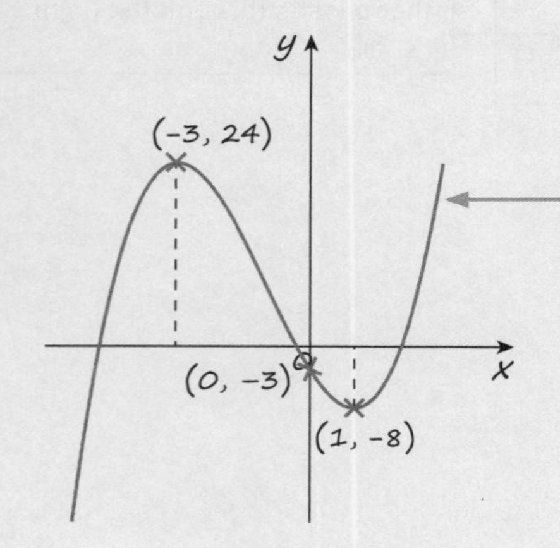

The coefficient of x^3 is positive (+1) and so the curve is this way up.

Test yourself

TESTED

1 Find $\int 2x^3\,dx$.

A $\frac{2x^3}{3}+c$ B $\frac{x^4}{2}+c$ C $6x^2+c$ D $\frac{2x^4}{3}+c$

2 Find $\int (3x^2+1)\,dx$.

A x^3+c B $6x+c$ C x^3+1+c D x^3+x+c

3 A curve has gradient given by $\frac{dy}{dx}=4x$ and passes through the point (–2, 3). What is the equation of the curve?

A $y=4x+11$ B $y=2x^2+11$ C $y=2x^2-20$ D $y=2x^2-5$

4 A curve has gradient given by $\frac{dy}{dx}=6x^2-6x$ and passes through the point (2, –1). What is the equation of the curve?

A $y=2x^3-3x^2-5$ B $y=12x-6$ C $y=6x^3-6x^2-23$ D $y=2x^3-3x^2+7$

5 A curve has gradient given by $\frac{dy}{dx}=x^2-2x+1$ and passes through the point (–1, –2). Where does the curve intercept the y-axis?

A $y=-1\frac{2}{3}$ B $y=-2$ C $y=-1$ D $y=\frac{1}{3}$

Full worked solutions online

CHECKED ANSWERS

Exam-style question

i Find $\int (6x^2+5x-3)\,dx$.

ii A curve has gradient given by $\frac{dy}{dx}=3x^2-2$. The curve passes through the point (2, –1). Find the equation of the curve.

Short answers on page 157

Full worked solutions online

CHECKED ANSWERS

Finding areas

REVISED

Key facts

1 A **definite integral** has **limits** which you substitute into the integrated function.

$$\int_a^b x^n\,dx=\left[\frac{x^{n+1}}{n+1}\right]_a^b=\frac{b^{n+1}}{n+1}-\frac{a^{n+1}}{n+1}$$

An **indefinite** integral has no limits.
$\int x^3\,dx$ is an example of an indefinite integral.

2

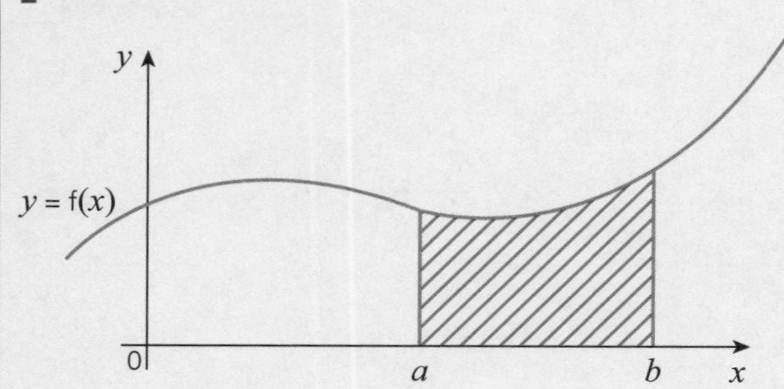

The shaded area between the curve $y = \mathrm{f}(x)$ and the x-axis, and between the values $x = a$ and $x = b$ is given by

$$A = \int_a^b \mathrm{f}(x)\mathrm{d}x = [\mathrm{F}(x)]_a^b = \mathrm{F}(b) - \mathrm{F}(a)$$

F(x) is obtained by integrating f(x).

3 Integration gives areas below the x-axis as negative.

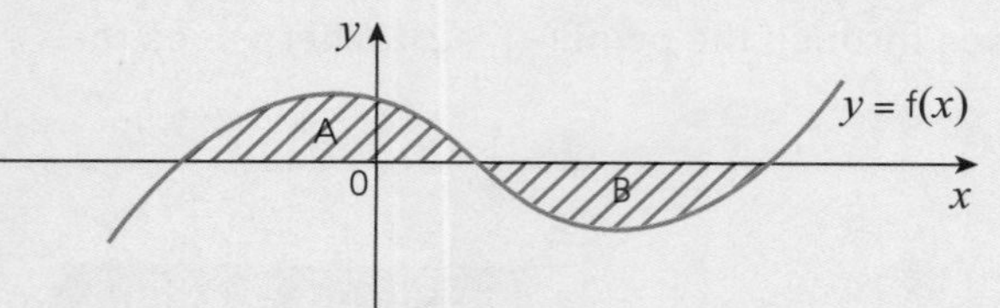

If $y = \mathrm{f}(x)$ crosses the x-axis in the required region then calculate the area in two parts, A and B, and add them together.
Total area is A + B.

Hint: Be careful when finding areas below the axis.
When you calculate the area of Region B the answer will have a negative sign. An area should always be given as positive.

Take the positive value of the area B.

Worked examples

1 Evaluating a definite integral

Evaluate the definite integral $\int_1^3 (x^4 + 4x^3 - 3x^2)\,\mathrm{d}x$.

Solution

$$\int_1^3 (x^4 + 4x^3 - 3x^2)\,dx = \left[\frac{x^5}{5} + \frac{4x^4}{4} - \frac{3x^3}{3}\right]_1^3$$

$$= \left[\frac{x^5}{5} + x^4 - x^3\right]_1^3$$

$$= \left(\frac{3^5}{5} + 3^4 - 3^3\right) - \left(\frac{1}{5} + 1 - 1\right)$$

$$= \frac{243}{5} + 81 - 27 - \frac{1}{5}$$

$$= \frac{242}{5} + 54 = \frac{512}{5} = 102.4$$

Remember $\int x^n \mathrm{d}x = \frac{x^{n+1}}{n+1} + c$
Add 1 to the power, and then divide by the new power.

Hint: If you are asked for an exact answer, do not use the 'numerical integration' function on your calculator. An exact answer may be an integer, or contain a fraction, a surd, or a number like π.

$\sqrt{3}$ is exact; 1.732 is not.

$\frac{29}{7}$ is exact; 4.143 is not.

2 Finding the area between a curve and the x-axis

For a region above the x-axis.

i Draw the graph of $y = (x+1)(2-x)$.

ii Find the area bounded by the curve $y = (x+1)(2-x)$ and the x-axis.

Solution

i *This curve crosses the x-axis at $x = -1$ and at $x = 2$ and the y-axis at 2*

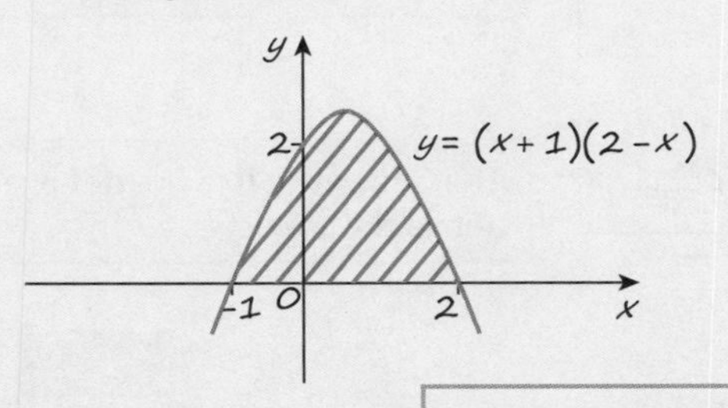

You cannot integrate this as it stands. You must multiply out the brackets first.

ii $\text{Area} = \int_{-1}^{2} (x+1)(2-x)\,dx = \int_{-1}^{2} (-x^2 + x + 2)\,dx$

$$= \left[\frac{-x^3}{3} + \frac{x^2}{2} + 2x\right]_{-1}^{2} = \left(\frac{-8}{3} + 2 + 4\right) - \left(\frac{1}{3} + \frac{1}{2} - 2\right)$$

$$= \frac{10}{3} - \left(-\frac{7}{6}\right) = \frac{9}{2} \text{ or } 4\frac{1}{2} \text{ square units}$$

Hint: Always sketch the curve to check whether it crosses the axis.

3 Finding areas below the x-axis

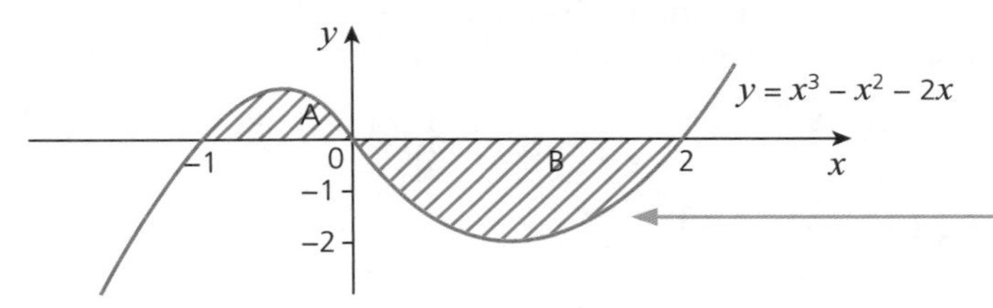

The graph shows the curve $y = x^3 - x^2 - 2x$.

i Find the area of the shaded region.

ii Evaluate $\int_{-1}^{2} (x^3 - x^2 - 2x)\,dx$.

iii Explain why your answers to parts **i** and **ii** are not the same.

Remember if an area is below the x-axis it has a negative sign. A curve may cross the x-axis within the area required. In this case evaluate the area above the axis and the area below the axis separately and then add them.

Solution

i $A = \int_{-1}^{0} (x^3 - x^2 - 2x)\,dx = \left[\frac{x^4}{4} - \frac{x^3}{3} - \frac{2x^2}{2}\right]_{-1}^{0}$

$$= (0) - \left(\frac{1}{4} + \frac{1}{3} - 1\right) = \frac{5}{12}$$

And $B = \int_{0}^{2} (x^3 - x^2 - 2x)\,dx = \left[\frac{x^4}{4} - \frac{x^3}{3} - \frac{2x^2}{2}\right]_{0}^{2}$

$$= \left(\frac{16}{4} - \frac{8}{3} - 4\right) - (0) = -\frac{8}{3}$$

This area is negative because it is below the x-axis.

Evaluate the area above the axis and the area below the axis separately and then add them.

Total shaded area $= A + B = \frac{5}{12} + \frac{8}{3} = \frac{37}{12}$ *or* $3\frac{1}{12}$ *square units.*

ii $\int_{-1}^{2}(x^3 - x^2 - 2x)\,dx = \left[\frac{x^4}{4} - \frac{x^3}{3} - \frac{2x^2}{2}\right]_{-1}^{2}$

$= \left(\frac{16}{4} - \frac{8}{3} - 4\right) - \left(\frac{1}{4} + \frac{1}{3} - 1\right)$

$= \frac{-8}{3} + \frac{5}{12} = \frac{-27}{12}$ or $-2\frac{1}{4}$

iii *The answers to i and ii are different because in i the fractions are added without regard to the sign, but in ii the minus sign means that the fractions were subtracted.*

Common mistakes: If you are just asked to evaluate a definite integral you don't need to worry about whether the curve is above or below the x-axis.

The answer to **ii** is the **net** area above the x-axis.

Test yourself

TESTED

1 Here is the graph of $y = 4 - x^2$:

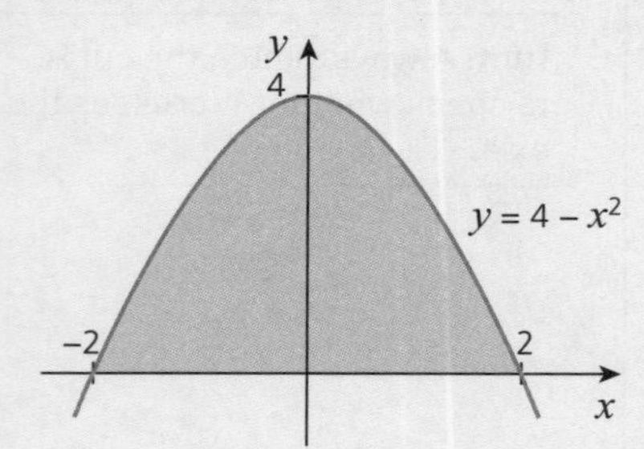

You are asked to find the area between the curve $y = 4 - x^2$ and the x-axis. Three of the following expressions will give the correct answer. Which of them **cannot** lead to the correct answer?

A $\int_{-2}^{2}(4 - x^2)\,dx$ B $\int_{0}^{4}(4 - x^2)\,dx$ C $\left[4x - \frac{x^3}{3}\right]_{-2}^{2}$ D $2\int_{0}^{2}(4 - x^2)\,dx$

2 The exact value of an area is to be found using $\left[\frac{x^2}{2} + \frac{x^3}{3} + \frac{x^4}{4}\right]_{1}^{2}$.

Which of the following is the correct answer?

A 7.58 B $2\frac{7}{9}$ C $7\frac{7}{12}$ D $9\frac{3}{4}$

3 Find the area enclosed between the curve $y = 6 + 2x - 3x^2$, the x-axis and the lines $x = 1$and $x = 1$.

A 2 B 10 C 12 D −12

Hint: First sketch the curve $y = 6 + 2x - 3x^2$ for $-1 \leqslant x \leqslant 1$ and the lines $x = 1$ and $x = -1$.

4 Here is the graph of $y = x^3 - x$:

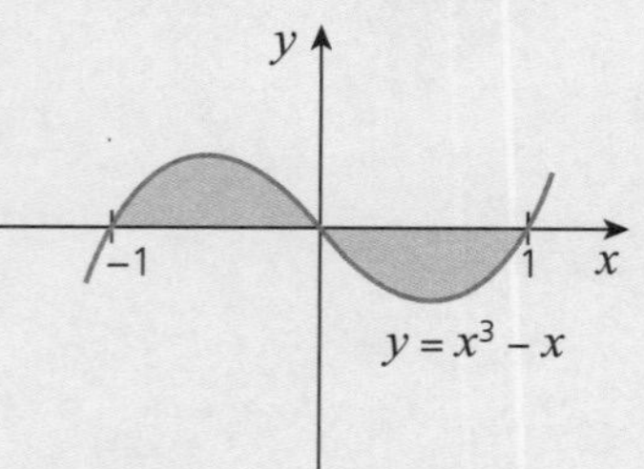

You are asked to find the shaded area between the curve $y = x^3 - x$ and the x-axis. One of the following expressions will give the correct answer. Which one leads to the correct answer?

A $-\int_{-1}^{0}(x^3 - x)dx + \int_{0}^{1}(x^3 - x)dx$ B $\left[\frac{x^4}{4} - \frac{x^2}{2}\right]_{-1}^{1}$ C $\int_{-1}^{0}(x^3 - x)dx + \int_{0}^{1}(x^3 - x)dx$ D $\int_{-1}^{0}(x^3 - x)dx - \int_{0}^{1}(x^3 - x)dx$

Full worked solutions online

CHECKED ANSWERS

Exam-style question

The diagram shows the white concrete facing of a tunnel.

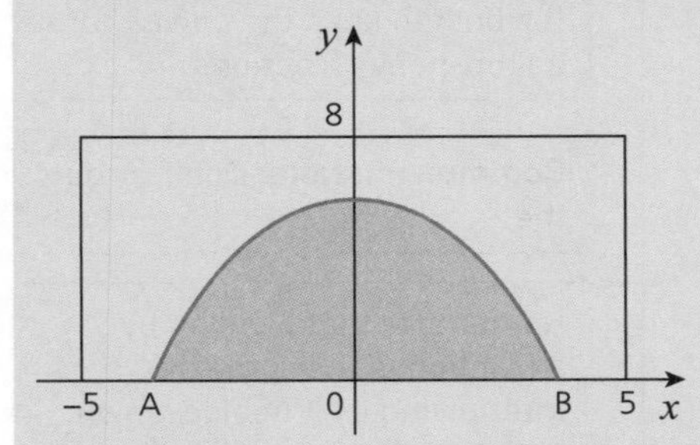

The x-axis represents the ground.

The roof of the tunnel has equation $y = 6 - \frac{3x^2}{8}$, where 1 unit is 1 metre.

i Find the coordinates of A and B.

ii Find the area between the curve and the x-axis.

iii Calculate the cost of repainting the facing at £5 per square metre.

Short answers on page 157

Full worked solutions online

CHECKED ANSWERS

Extending the rule

REVISED

Key fact

The **rule** for integrating a power of x:

$$\int ax^n \, dx = \frac{ax^{n+1}}{n+1} + c$$

can be used when n is a fraction or a negative number (except for –1).

Worked examples

1 Finding indefinite integrals

Find:

i $\int \frac{8}{x^3} dx$ ii $\int 3\sqrt{x} \, dx$ iii $\int \left(x^2 + \frac{1}{x^2}\right) dx$ iv $\int \frac{\sqrt{x}}{x} dx.$

Solution

i $\int \frac{8}{x^3} dx = \int 8x^{-3} dx$

$= \frac{8x^{-2}}{-2} + c$

$= -\frac{4}{x^2} + c$

Rewrite as a single power of x. Remember $\frac{1}{x^n} = x^{-n}$

Common mistake: Be careful when adding 1 to a negative number!

It's usual to rewrite the answer in the same form as the question.

ii $\int 3\sqrt{x}\,dx = \int 3x^{\frac{1}{2}}\,dx$

$$= \frac{3x^{\frac{3}{2}}}{\frac{3}{2}} + c$$

$$= 2\sqrt{x^3} + c$$

iii $\int \left(x^2 + \frac{1}{x^2}\right)dx = \int \left(x^2 + x^{-2}\right)dx$

$$= \frac{x^3}{3} + \frac{x^{-1}}{-1} + c$$

$$= \frac{x^3}{3} - \frac{1}{x} + c$$

iv $\int \frac{\sqrt{x}}{x}\,dx = \int \frac{x^{\frac{1}{2}}}{x^1}\,dx$

$$= \int x^{-\frac{1}{2}}\,dx$$

$$= \frac{x^{\frac{1}{2}}}{\frac{1}{2}} + c$$

$$= 2\sqrt{x} + c$$

Roots of x can be written as fractional powers.

It's best to keep the powers of x as top-heavy fractions.

Common mistake: Don't forget the + c!

Remember that dividing by a fraction is the same as multiplying by it upside-down.

Hint: You can't integrate this as it stands, you need to write it as a single power of x.

Remember the laws of indices:

$$\frac{x^a}{x^b} = x^{a-b}$$

2 Evaluating definite integrals

i Find $\int_2^5 \frac{3}{x^2}\,dx$

ii Find $\int_1^8 \sqrt[3]{x}\,dx$.

Solution

i $\int_2^5 \frac{3}{x^2}\,dx = \int_2^5 3x^{-2}\,dx$

$$= \left[\frac{3x^{-1}}{-1}\right]_2^5$$

$$= \left[-\frac{3}{x}\right]_2^5$$

$$= \left(-\frac{3}{5}\right) - \left(-\frac{3}{2}\right)$$

$$= \frac{9}{10}$$

Hint: You might find it easier to rewrite x^{-1} as $\frac{1}{x}$ to help you when evaluating it at $x = 5$ and $x = 2$.

$$
\textbf{ii} \quad \int_1^8 \sqrt[3]{x}\,dx = \int_1^8 x^{\frac{1}{3}}\,dx
$$

$$
= \left[\frac{x^{\frac{4}{3}}}{\frac{4}{3}}\right]_1^8
$$

$$
= \left[\frac{3}{4}\left(\sqrt[3]{x}\right)^4\right]_1^8
$$

$$
= \left[\frac{3}{4}\left(\sqrt[3]{8}\right)^4\right] - \left[\frac{3}{4}\left(\sqrt[3]{1}\right)^4\right]
$$

$$
= \left(\frac{3}{4}\times 16\right) - \left(\frac{3}{4}\times 1\right)
$$

$$
= 11\frac{1}{4}
$$

Hint: Remember what fractional indices mean. $x^{\frac{4}{3}}$ is the same as $\left(\sqrt[3]{x}\right)^4$.
You can also write it as $\sqrt[3]{x^4}$ but it is often easier to work out the numbers in the form used here.

3 Finding the area under a curve

The graph of $y = x^2 + \frac{1}{x^3}$ is shown.

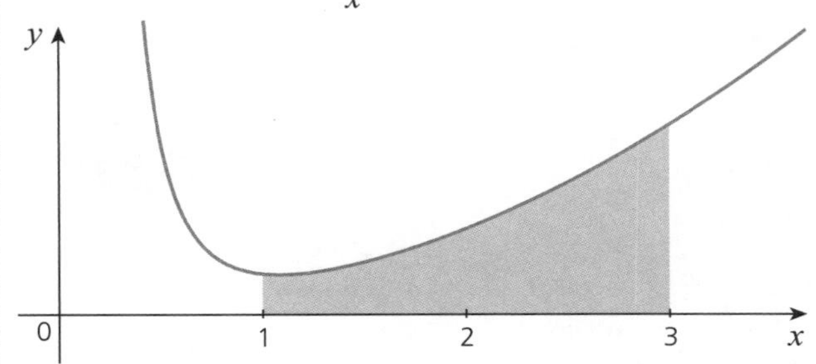

The shaded region is bounded by the curve, the x-axis and the lines $x = 1$ and $x = 3$.
Find the area of the shaded region.

Solution

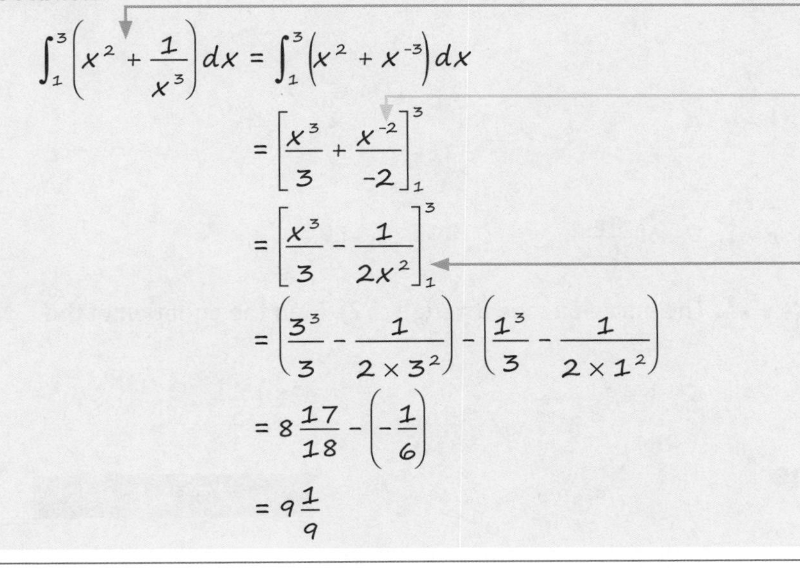

$$
\int_1^3 \left(x^2 + \frac{1}{x^3}\right)dx = \int_1^3 \left(x^2 + x^{-3}\right)dx
$$

$$
= \left[\frac{x^3}{3} + \frac{x^{-2}}{-2}\right]_1^3
$$

$$
= \left[\frac{x^3}{3} - \frac{1}{2x^2}\right]_1^3
$$

$$
= \left(\frac{3^3}{3} - \frac{1}{2\times 3^2}\right) - \left(\frac{1^3}{3} - \frac{1}{2\times 1^2}\right)
$$

$$
= 8\frac{17}{18} - \left(-\frac{1}{6}\right)
$$

$$
= 9\frac{1}{9}
$$

To find an area integrate the function between the two limits.

Common mistake: Be careful when adding 1 to a negative number!

Rewriting should help you find the values at $x = 3$ and 1.

Chapter 11 Integration

4 Finding the equation of a curve given its gradient function, $\frac{dy}{dx}$

The gradient of a curve is given by $\frac{dy}{dx} = \frac{5}{x^4}$. The curve passes through (1, 2). Find the equation of the curve.

Solution

$\frac{dy}{dx} = \frac{5}{x^4}$ can be rewritten as $\frac{dy}{dx} = 5x^{-4}$

$$y = \int 5x^{-4}\,dx$$

$$= \frac{5x^{-3}}{-3} + c$$

$$= -\frac{5}{3x^3} + c$$

Substitute $x = 1$ and $y = 2$ into the equation:

$$2 = -\frac{5}{3 \times 1^3} + c, \text{ so } c = 2 + \frac{5}{3} = \frac{11}{3}$$

So the equation of the curve is $y = \frac{11}{3} - \frac{5}{3x^3}$

If you know $\frac{dy}{dx}$ then integrating will give you y.

Use the coordinates of the point you have been given to find the value of c.

Test yourself

TESTED

1 Find $\int \frac{3}{x^2}\,dx$.

A $3x^{-2} + c$ B $-\frac{6}{x^3} + c$ C $-\frac{3}{x} + c$ D $2x^{\frac{3}{2}} + c$

2 Find $\int \sqrt[3]{x}\,dx$.

A $-\frac{1}{2x^2} + c$ B $x^{\frac{1}{3}} + c$ C $\frac{1}{3\sqrt[3]{x^2}} + c$ D $\frac{3(\sqrt[3]{x})^4}{4} + c$

3 Find $\int_1^4 \sqrt{x}\,dx$.

A $\frac{21}{2}$ B $\frac{3}{4}$ C 1 D $\frac{14}{3}$

4 Find $\int_2^4 \left(x^3 + \frac{1}{x^3}\right)dx$.

A $56\frac{9}{64}$ B $60\frac{3}{32}$ C $60\frac{15}{1024}$ D $59\frac{29}{32}$

5 The gradient of a curve is given by $\frac{dy}{dx} = \frac{3}{x^4}$. The curve passes through (1, 2). Find the equation of the curve.

A $y = 3 - \frac{1}{x^3}$ B $y = -\frac{12}{x^5}$ C $y = 1\frac{1}{8} - \frac{1}{x^3}$ D $y = 2\frac{3}{5} - \frac{3}{5x^5}$

Full worked solutions online

CHECKED ANSWERS

Exam-style question

i Find $\int \left(3\sqrt{x} + \frac{1}{x^2}\right) \mathrm{d}x$.

ii A curve has a gradient given by $\frac{\mathrm{d}y}{\mathrm{d}x} = \frac{2}{x^2}$. The curve passes through the point (3, 1). Find the equation of the curve.

Short answers on page 157

Full worked solutions online

CHECKED ANSWERS

Chapter 12 Vectors

About this topic

Vectors are important in many areas of mathematics and you will also use them in mechanics.

Before you start, remember

- Pythagoras' theorem
- coordinate geometry.

Working with vectors

REVISED

Key facts

1. A **scalar** quantity has **magnitude** only.
2. A **vector** quantity has both **magnitude** and **direction**.
3. A **unit vector** is a vector with magnitude 1.
4. Vectors are printed in bold, e.g. **a**. In handwriting, vectors are underlined, e.g. $\underset{\sim}{a}$.
   The vector from point A to point B is written as $\overrightarrow{AB}$.
5. You can describe vectors using **magnitude-direction form.**
   A vector in two dimensions is often represented by a straight line with an arrowhead, which shows the direction of the vector.
   The direction is usually taken to be the angle that the vector makes with the positive x-axis, measured in an anticlockwise direction.
   The length of the line represents the magnitude of the vector.
6. You can also describe vectors using **components.**
   Vectors can be described using the unit vectors **i** and **j** in the x and y directions respectively. Vectors can also be written as column vectors.
7. Two vectors are **equal** if they have the same magnitude and the same direction.
8. Two vectors are **parallel** if they have the same direction.
9. The **position vector** of a point P is the vector from the origin, O, to P. This is written as $\overrightarrow{OP}$ or **p**.
   The point P(a, b) has position vector $\overrightarrow{OP} = a\mathbf{i} + b\mathbf{j}$, or as a **column vector**, $\begin{pmatrix} a \\ b \end{pmatrix}$.

20°
8

j
i

Magnitude means size.
Mass (kg) and speed (m s^{-1}) are scalar quantities.

Weight (N) and velocity (m s^{-1}) are vector quantities.

This vector has magnitude 8 units and direction –20° to the positive x-axis. The direction can also be given as 340°.

The vector shown in red can be written as $-5\mathbf{i} + 2\mathbf{j}$, or as the column vector $\begin{pmatrix} -5 \\ 2 \end{pmatrix}$.

The origin is not always shown in diagrams.

10 The **magnitude** of a vector is found using Pythagoras' theorem.

The vector $a\mathbf{i} + b\mathbf{j}$. has magnitude $\sqrt{a^2 + b^2}$.

The magnitude of the vector $\overrightarrow{\text{OP}}$ is written $|\overrightarrow{\text{OP}}|$.

The **direction** of $\overrightarrow{\text{OP}} = a\mathbf{i} + b\mathbf{j}$ is $\tan^{-1}\frac{b}{a}$.

11 The **unit vector** in the direction of $a\mathbf{i} + b\mathbf{j}$ is $\frac{a}{\sqrt{a^2+b^2}}\mathbf{i} + \frac{b}{\sqrt{a^2+b^2}}\mathbf{j}$.

12 To find the **resultant** of two or more vectors you add the vectors. This is particularly useful in mechanics for finding the resultant of two or more forces.

The magnitude of the vector $-5\mathbf{i} + 2\mathbf{j}$ is $\sqrt{(-5)^2 + 2^2} = \sqrt{25+4} = \sqrt{29}$.

The resultant of the vectors **a**, **b** and **c** is **a** + **b** + **c**

Worked examples

1 Calculating with vectors

Two vectors are given by $\mathbf{a} = \begin{pmatrix} 3 \\ 1 \end{pmatrix}$ and $\mathbf{b} = \begin{pmatrix} -1 \\ 2 \end{pmatrix}$.

Find the vectors: **i** 2**a** **ii** **a** + **b** **iii** **a** − **b**.

Hint: When a vector is multiplied by a scalar (a number), each component is multiplied by the scalar.

Two vectors in component form can be added or subtracted by dealing with each component separately.

Solution

i $2\underline{a} = 2\begin{pmatrix} 3 \\ 1 \end{pmatrix} = \begin{pmatrix} 6 \\ 2 \end{pmatrix}$

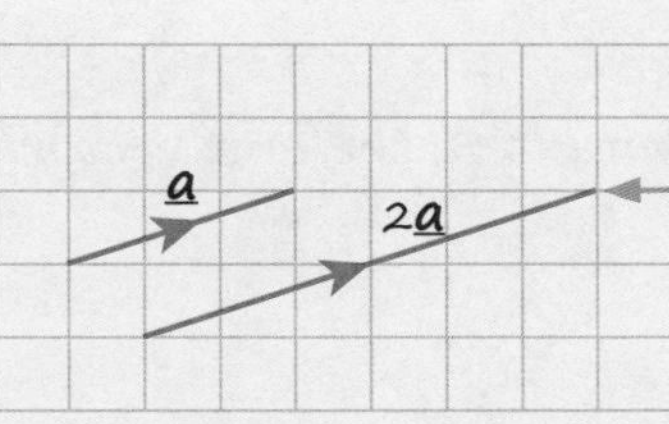

The vector 2$\underline{\mathbf{a}}$ is twice as long as $\underline{\mathbf{a}}$, in the same direction.

ii

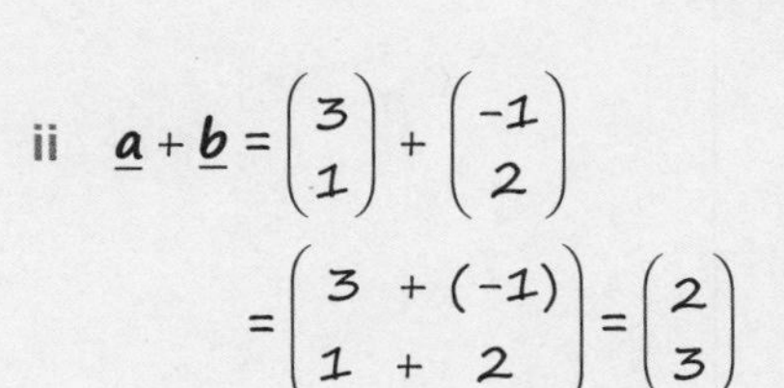

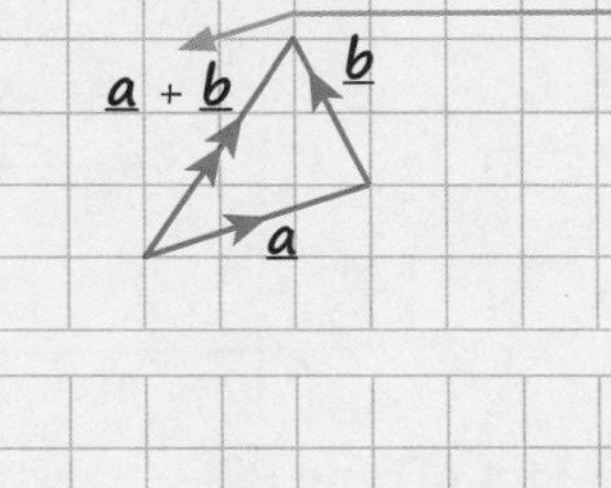

The vector $\underline{\mathbf{a}}$ + $\underline{\mathbf{b}}$ is equivalent to vector $\underline{\mathbf{a}}$ followed by vector $\underline{\mathbf{b}}$. This is the resultant of $\underline{\mathbf{a}}$ and $\underline{\mathbf{b}}$.

iii

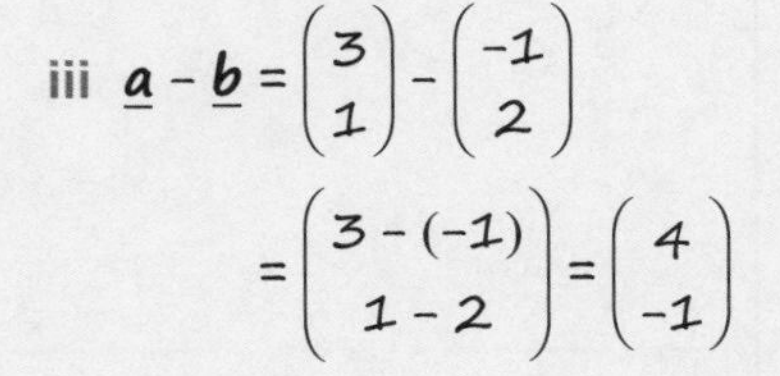

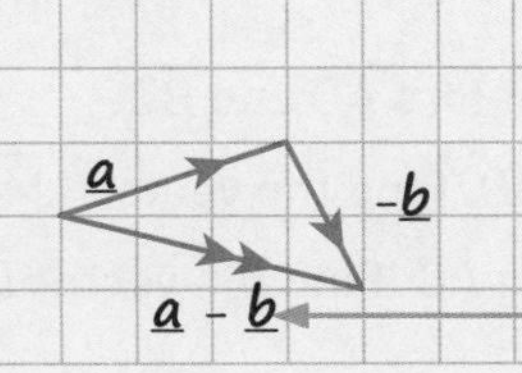

The vector $\underline{\mathbf{a}}$ − $\underline{\mathbf{b}}$ is equivalent to vector $\underline{\mathbf{a}}$ followed by vector −$\underline{\mathbf{b}}$, which is in the opposite direction to $\underline{\mathbf{b}}$.

2 Finding the magnitude of a vector and using unit vectors

The points A and B have coordinates (5, –1) and (2, 3).

i Find the vector $\overrightarrow{AB}$.

ii Find the magnitude and direction of the vector $\overrightarrow{AB}$.

iii Write down a unit vector in the direction of the vector $\overrightarrow{BA}$.

Solution

i $\overrightarrow{AB} = \overrightarrow{AO} + \overrightarrow{OB} = -\overrightarrow{OA} + \overrightarrow{OB}$

$$= -\begin{pmatrix} 5 \\ -1 \end{pmatrix} + \begin{pmatrix} 2 \\ 3 \end{pmatrix} = \begin{pmatrix} -5+2 \\ -(-1)+3 \end{pmatrix} = \begin{pmatrix} -3 \\ 4 \end{pmatrix}$$

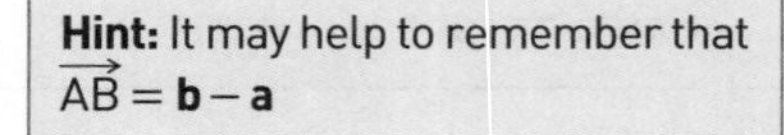

Hint: It may help to remember that $\overrightarrow{AB} = \mathbf{b} - \mathbf{a}$

ii $|\overrightarrow{AB}| = \sqrt{(-3)^2 + 4^2} = \sqrt{9+16} = \sqrt{25} = 5$

Direction of $\overrightarrow{AB} = \tan^{-1}\left(\frac{4}{-3}\right) = -53.1°$

So the direction is $180° - 53.1° = 126.9°$ to the positive $\mathbf{i}$ direction

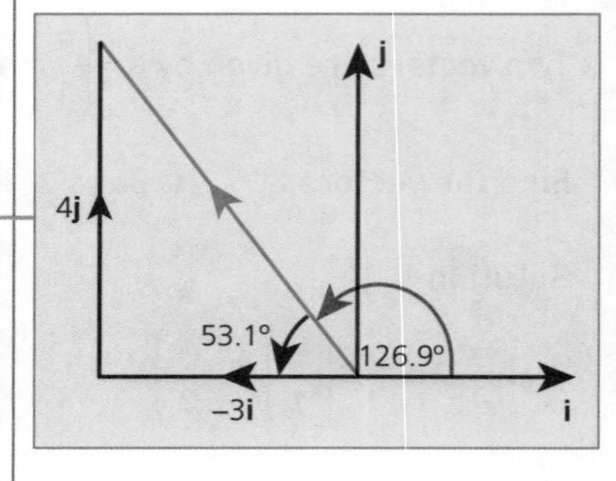

iii $\overrightarrow{BA} = -\overrightarrow{AB} = (3 \ \ -4)$

Since this vector has magnitude 5, the unit vector in this direction is $\frac{1}{5}\begin{pmatrix} 3 \\ -4 \end{pmatrix} = \begin{pmatrix} \frac{3}{5} \\ -\frac{4}{5} \end{pmatrix}$.

3 Geometry using vectors

The diagram shows a parallelogram ABCD.

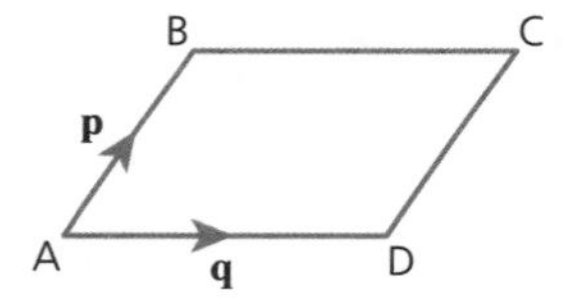

$\overrightarrow{AB} = \vec{\mathbf{p}}$ and $\overrightarrow{AD} = \vec{\mathbf{q}}$.

i Find, in terms of **p** and **q**, the vectors $\overrightarrow{AC}$ and $\overrightarrow{BD}$.

ii The point M is the midpoint of BD. Find the vector $\overrightarrow{AM}$.

iii The point N is $\frac{1}{3}$ of the way along AC. Find the vector $\overrightarrow{DN}$.

Solution

i $\overrightarrow{AC} = \overrightarrow{AB} + \overrightarrow{BC} = \underline{p} + \underline{q}$

As ABCD is a parallelogram, $\overrightarrow{BC} = \overrightarrow{AD} = \mathbf{q}$.

$\overrightarrow{BD} = \overrightarrow{BA} + \overrightarrow{AD} = -\underline{p} + \underline{q} = \underline{q} - \underline{p}$

ii $\overrightarrow{BM} = \frac{1}{2}\overrightarrow{BD} = \frac{1}{2}(\underline{q} - \underline{p})$

$\overrightarrow{AM} = \overrightarrow{AB} + \overrightarrow{BM} = \underline{p} + \frac{1}{2}(\underline{q} - \underline{p}) = \underline{p} + \frac{1}{2}\underline{q} - \frac{1}{2}\underline{p} = \frac{1}{2}(\underline{p} + \underline{q})$

iii $\overrightarrow{AN} = \frac{1}{3}\overrightarrow{AC} = \frac{1}{3}(\underline{p} + \underline{q})$

$\overrightarrow{DN} = \overrightarrow{DA} + \overrightarrow{AN} = -\underline{q} + \frac{1}{3}(\underline{p} + \underline{q}) = -\underline{q} + \frac{1}{3}\underline{p} + \frac{1}{3}\underline{q} = \frac{1}{3}\underline{p} - \frac{2}{3}\underline{q}$

Test yourself

TESTED

1 The vectors **a** and **b** are given by $\mathbf{a} = -\mathbf{j}$ and $\mathbf{b} = 3\mathbf{i} - 2\mathbf{j}$.

Find the vector $2\mathbf{a} - 3\mathbf{b}$.

A $-9\mathbf{i} - 5\mathbf{j}$ B $-9\mathbf{i} - 8\mathbf{j}$ C $-9\mathbf{i} + 4\mathbf{j}$ D $-11\mathbf{i} + 6\mathbf{j}$

Questions 2 to 5 are about three points A, B and C with position vectors $\begin{pmatrix}1\\0\end{pmatrix}$, $\begin{pmatrix}3\\1\end{pmatrix}$ and $\begin{pmatrix}-2\\6\end{pmatrix}$ respectively.

2 Find the vector $\overrightarrow{BA}$.

A $\begin{pmatrix}2\\1\end{pmatrix}$ B $\begin{pmatrix}4\\1\end{pmatrix}$ C $\begin{pmatrix}3\\0\end{pmatrix}$ D $\begin{pmatrix}-2\\-1\end{pmatrix}$

3 Find the unit vector in the direction $\overrightarrow{AC}$.

A $\begin{pmatrix}-3\\6\end{pmatrix}$ B $\begin{pmatrix}-\frac{\sqrt{5}}{5}\\\frac{2\sqrt{5}}{5}\end{pmatrix}$ C $\begin{pmatrix}-1\\2\end{pmatrix}$ D $\begin{pmatrix}-\frac{\sqrt{3}}{3}\\\frac{2\sqrt{3}}{3}\end{pmatrix}$

4 Find the position vector of the point D so that ABCD is a parallelogram.

A $\begin{pmatrix}-5\\5\end{pmatrix}$ B $1\begin{pmatrix}6\\-5\end{pmatrix}$ C $\begin{pmatrix}0\\7\end{pmatrix}$ D $\begin{pmatrix}-4\\5\end{pmatrix}$

5 Find the position vector of the midpoint M of BC.

A $\begin{pmatrix}0.5\\3.5\end{pmatrix}$ B $\begin{pmatrix}-2.5\\2.5\end{pmatrix}$ C $\begin{pmatrix}2\\4\end{pmatrix}$ D $\begin{pmatrix}-4.5\\8.5\end{pmatrix}$

Full worked solutions online

CHECKED ANSWERS

Exam-style question

The points A, B and C have position vectors $2\mathbf{i} + 3\mathbf{j}$, $4\mathbf{i} - 5\mathbf{j}$ and $-\mathbf{i} + 4\mathbf{j}$ respectively.

i Find the vector $\overrightarrow{AB}$.

ii Find the magnitude and direction of $\overrightarrow{AB}$, where the direction is measured from the positive **i** direction.

iii Find the position vector of M when:

a $\overrightarrow{AB} = \overrightarrow{CM}$ b M is the midpoint of $\overrightarrow{AB}$.

Short answers on page 157

Full worked solutions online

CHECKED ANSWERS

Chapter 13 Exponentials and logarithms

About this topic

This topic looks at exponential functions, and their inverses – logarithms. The most common logarithms are to base 10 ($\log_{10}$) and to base e where e is the irrational number 2.718...). Exponential functions have many real-life applications, for example in modelling population growth or radioactive decay.

Before you start, remember

- the laws of indices.

Exponential functions and logarithms

REVISED

Key facts

1 $f(x) = a^x$ is an exponential function. An **exponential function** is a function which has the variable as the power, such as 2^x. (An alternative name for power is **exponent**).

2 Exponential functions model real-life situations and all graphs follow a similar pattern to one of the curves below.

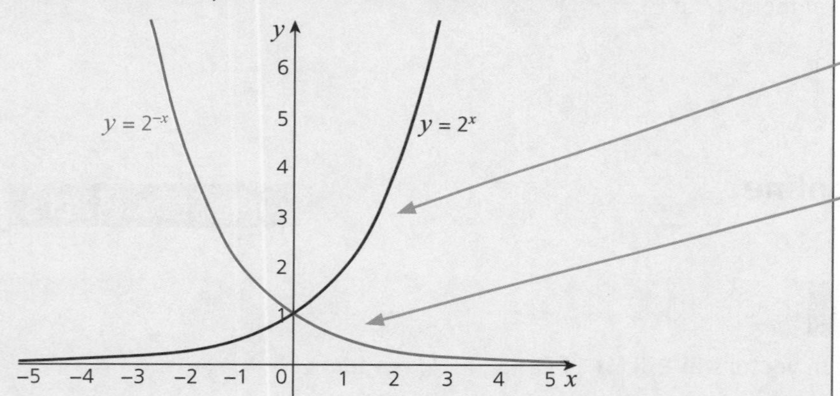

3 **Logarithm** is another word for **index** or **power**.
A logarithm is the **inverse** of the exponential function.
So $y = \log_2 x$ is the inverse function of $y = 2^x$.
This is true for any base a, not just 2;

$y = \log_a x \Leftrightarrow a^y = x$

$\log_a(a^x) = x$ and $a^{\log_a x} = x$

The graph of $y = a^x$ with $a > 1$, has a gradient that is increasing. This is described as **exponential growth**.

The graph of $y = a^{-x}$ with $a > 1$, has a gradient that is decreasing. This is described as **exponential decay**.

Hint: Notice that both curves:
- pass through (0, 1)
- lie above the x-axis (so y is positive for all values of x)
- have a gradient that is proportional to the y coordinate, so $\frac{dy}{dx} \propto y$.

4 The graph of $y = \log_a x$ is the reflection of $y = a^x$ in the line $y = x$.

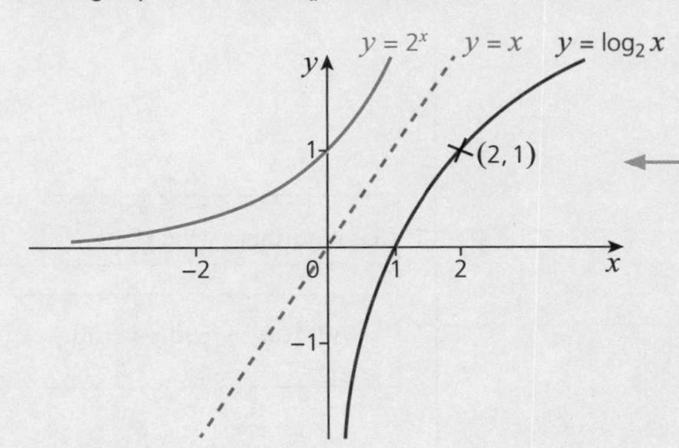

The graph of $y = \log_a x$ looks like $y = \log_2 x$. It goes through (1, 0) and $(a, 1)$.

Notice that for values of x between 0 and 1, $\log x$ is negative.

$\log x$ is only defined for positive values of x, so log (−2) is undefined.

5 For logarithms to any base:

- Multiplication: $\log xy = \log x + \log y$
- Division: $\log \frac{x}{y} = \log x - \log y$
- Powers: $\log x^n = n \log x$
- Roots: $\log \sqrt[n]{x} = \frac{1}{n} \log x$
- Logarithm of 1: $\log 1 = 0$
- Reciprocals: $\log \frac{1}{x} = -\log x$
- Logarithm to its own base: $\log_a a = 1$

If a logarithmic expression is true for any base, then the base will often be omitted.

Hint: There may be two or three logarithm buttons on your calculator, depending on how sophisticated it is. For this section, you should use the basic log button which may be labelled just 'log' or '$\log_{10}$'.

6 Logarithms can be used to solve equations involving powers to any level of accuracy.

$$2^x = 100$$
$$\Rightarrow \quad \log 2^x = \log 100$$
$$\Rightarrow \quad x \log 2 = \log 100$$
$$\Rightarrow x = \frac{\log 100}{\log 2} = 6.64 \text{ (3 s.f.)}$$

Worked examples

1 Evaluating logarithms

Find the values of the following without using a calculator:

i $\log_3 81$ **ii** $\log_5\left(\frac{1}{25}\right)$ **iii** $\log_4 2$

Solution

i $\log_3 81 = \log_3 3^4 = 4\log_3 3 = 4$

Write 81 as a power of 3 and then use $\log x^n = n \log x$.

ii $\log_5\left(\frac{1}{25}\right) = \log_5 1 - \log_5 25 = 0 - \log_5 5^2 = -2\log_5 5 = -2$

or $\frac{1}{25} = 5^{-2}$, so $\log 5 = -2$

Use $\log \frac{x}{y} = \log x - \log y$.

iii $\log_4 2 = \log_4 \sqrt{4} = \log_4 4^{\frac{1}{2}} = \frac{1}{2}\log_4 4 = \frac{1}{2}$

Use $\log_a a = 1$.

2 Manipulating equations containing logarithms

Given that $\log y = \log(x+3) - 3\log 2$, find y in terms of x.

Solution

$$\log y = \log(x+3) - 3\log 2$$
$$= \log(x+3) - \log 2^3$$
$$= \log(x+3) - \log 8$$
$$= \log \frac{x+3}{8}$$
$$\Rightarrow y = \frac{x+3}{8}$$

Using $n\log x = \log x^n$.

Using $\log x - \log y = \log\frac{x}{y}$.

3 Solving equations containing logarithms

i Write $2\log x - \log 11$ as a single logarithm.

ii Hence solve the equation $2\log x - \log 11 = \log 44$.

Solution

i $2\log x - \log 11 = \log x^2 - \log 11 = \log \frac{x^2}{11}$

ii $2\log x - \log 11 = \log 44$

$$\Rightarrow \log\frac{x^2}{11} = \log 44$$
$$\Rightarrow \frac{x^2}{11} = 44$$
$$\Rightarrow x^2 = 484$$
$$\Rightarrow x = 22$$

Using $n\log x = \log x^r$.

Using $\log x - \log y = \log\frac{x}{y}$.

Common mistake: The option $x = -22$ has been rejected since the original equation contained $\log x$ and this is only defined for positive values of x.

4 Using logarithms to evaluate a power

Solve:

i $5^x = 3000$

ii $5^{2x} - 6 \times 5^x + 5 = 0$.

Solution

i
$$\log 5^x = \log 3000$$
$$\Rightarrow x\log 5 = \log 3000$$
$$\Rightarrow x = \frac{\log 3000}{\log 5}$$
$$\Rightarrow x = 4.97 \text{ (3 s.f.)}$$

Take logs of both sides.

ii Rewrite $5^{2x} - 6 \times 5^x + 5 = 0$ as $(5^x)^2 - 6 \times 5^x + 5 = 0$

Let $y = 5^x \Rightarrow y^2 - 6y + 5 = 0$

$$(y-5)(y-1) = 0$$
$$\Rightarrow y = 5 \text{ or } y = 1$$
$$\Rightarrow 5^x = 5 \text{ or } 5^x = 1$$
$$\Rightarrow x = 1 \text{ or } x = 0$$

This is a quadratic equation in disguise.

You can solve these by inspection or you can take logs of both sides like in part i.

5 Modelling exponential growth

Following the decision by a major international company to set up its base in the UK, a new town is being developed and it is estimated that the growth of population over the first five years will be modelled by the equation $P = 15\,000 \times 10^{0.15t}$ where t is the time in years.

i Calculate the population at the end of the first year, giving your answer to 3 s.f.

ii When will the population exceed 50 000?

iii Why is this not a suitable model in the long term?

Solution

i After 1 year: $P = 15\,000 \times 10^{0.15 \times 1} = 21\,200$

Substitute $t = 1$ into $P = 15\,000 \times 10^{0.15t}$.

ii Solve the inequality $15\,000 \times 10^{0.15t} > 50\,000$

$$10^{0.15t} > \frac{50\,000}{15\,000}$$

$$\log 10^{0.15t} > \log\left(\frac{50\,000}{15\,000}\right)$$

Taking logs of both sides.

$$0.15t \log 10 > \log\left(\frac{10}{3}\right)$$

Simplify the fraction.

$$0.15t > \log\left(\frac{10}{3}\right)$$

$$t > 3.48\ldots$$

$$\Rightarrow t = 4$$

The population will exceed 50 000 during the fourth year.

iii After 20 years the model predicts the population of the town would be $P = 15\,000 \times 10^3 = 15\,000\,000$ which is unrealistic.

It is likely that the population will increase exponentially for a few years and then gradually stabilise.

Hint: It is a good idea to check your answer.
After 3 years,
$P = 15\,000 \times 10^{0.45} = 42\,300$.
After 4 years,
$P = 15\,000 \times 10^{0.60} = 59\,700$.

Test yourself

TESTED

1 Which of the following is the correct answer when writing $\log 12 - 3\log 2 + 2\log 3$ as a single logarithm?

A $\log 12$ B $\log 13.5$ C $\log\frac{1}{6}$ D $\log 10\frac{2}{3}$

2 Use logarithms to base 10 to solve the equation $2.5^x = 1000$ to 2 d.p.

A 2.90 B 3.00 C 7.54 D 7.53882

3 Simplify $\frac{1}{2}\log 64 - 2\log 2$ writing your answer in the form $\log x$.

A $\log 8$ B $\log 2$ C $\log 32$ D 0.301

4 Express $\log\sqrt{x} + \log x^{\frac{7}{2}} - 2\log x$ as a single logarithm.

A $2\log x$ B $\log x^2$ C $\log\left(x^{-\frac{3}{2}} + x^{\frac{3}{2}}\right)$ D $\log\left(\frac{x^{\frac{1}{2}} \times x^{\frac{7}{2}}}{x^2}\right)$

5 The value of an investment varies according to the formula $V = Ae^{0.1t}$, where e = 2.71828 to 5 d.p. and t is the period of investment in years.
The investment is predicted to be worth £10 000 after 5 years.
Find the value of A to the nearest £.

A £16 487 B £6 065 C £7 358 D £6 065.31

Full worked solutions online

CHECKED ANSWERS

Exam-style question

i a Express $\log_a \frac{1}{p^2} + \log_a p^5$ as a multiple of $\log_a p$.

b Hence solve the equations:

A $\log_{10} \frac{1}{p^2} + \log_{10} p^5 = 9$

B $\log_2 \frac{1}{p^2} + \log_2 p^5 = 9$

ii Solve the equation $\log z + \log(z-2) = \log 3$.

Short answers on page 157

Full worked solutions online

CHECKED ANSWERS

Natural logarithms and exponentials

REVISED

Key facts

1 e is the number 2.718281...

2 e^x is the **exponential function**.
Exponential functions obey the usual rules of indices:

$e^x \times e^y = e^{x+y}$ $e^x \div e^y = e^{x-y}$
$e^1 = e$ $e^0 = 1$ $e^{-1} = \frac{1}{e}$

3 **Exponential growth and decay**

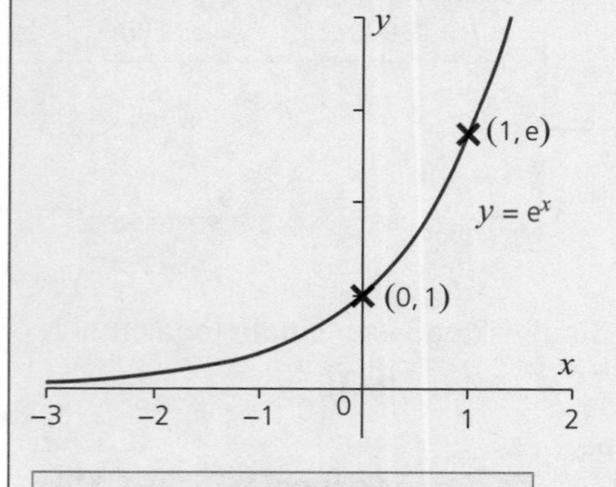

y
(−1, e)
$y = e^{-x}$
1
(0, 1)
x
−2 −1 0 1 2 3

The graph of $y = e^x$ shows **exponential growth**.
As $x \to \infty, y \to \infty$ at an increasing rate.
The gradient at any point equals the y coordinate, so $\frac{dy}{dx} = e^x$

The graph of $y = e^{-x}$ shows **exponential decay**.
As $x \to \infty, y \to 0$ at a decreasing rate.
The gradient at any point equals $-y$, so $\frac{dy}{dx} = -e^{-x}$

4 $y = e^{kx} \Rightarrow \frac{dy}{dx} = ke^{kx}$

5 The natural logarithm of x is written as $\ln x$ or $\log_e x$
e^x and $\ln x$ are inverse functions.

$y = \ln x \Leftrightarrow x = e^y$

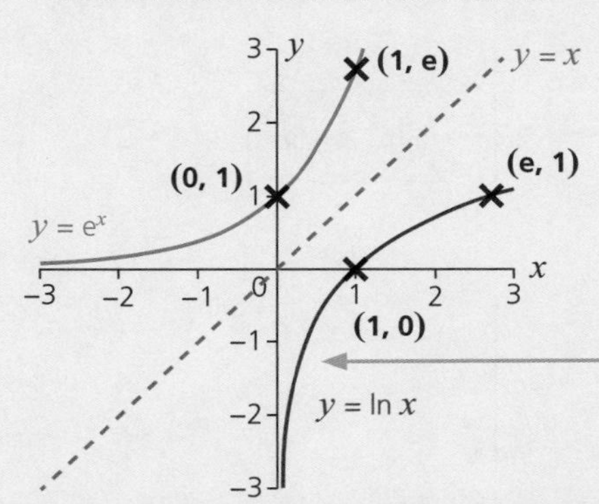

The negative x-axis is an asymptote for $y = e^x$

The negative y-axis is an asymptote for $y = \ln x$

Notice that for values of x between 0 and 1, $\ln x$ is negative.

6 Natural logarithms obey the same rules as logs to the base 10, or any other base.

$\ln(a \times b) = \ln a + \ln b$ $\quad \ln(a \div b) = \ln a - \ln b$ $\quad \ln\left(\frac{1}{a}\right) = -\ln a$

$\ln(a^n) = n\ln a$ $\quad \ln(\sqrt[n]{a}) = \frac{1}{a}\ln a$

$\ln 1 = 0$ $\quad \ln e = 1$

Worked examples

1 Differentiating the exponential function

Give $f(x) = 20x - 3e^{5x}$, find $f'(0)$.

Solution

$$f(x) = 20x - 3e^{5x} \Rightarrow f'(x) = 20 - 3 \times 5e^{5x}$$
$$\Rightarrow f'(x) = 20 - 15e^{5x}$$
$$\Rightarrow f'(0) = 20 - 15e^{0}$$
$$\Rightarrow \quad = 20 - 15$$
$$\Rightarrow \quad = 5$$

Using $y = e^{kx} \Rightarrow \frac{dy}{dx} = ke^{kx}$

Remember any number to the power of 0 is 1.

2 Switching between natural logarithms and exponentials

Solve the equation $e^{2x} = 3$.

Solution

Take ln of both sides $\ln(e^{2x}) = \ln 3$
$$2x = \ln 3$$
$$x = \frac{\ln 3}{2} = 0.549 \text{ to 3 d.p.}$$

Remember $\ln(e^x) = x$ and $e^{\ln x} = x$.

Common mistake: You can only have the logarithm of a positive number. So if you get, say, $\ln(-3)$ something has gone wrong. You have almost certainly made a mistake and should check back.

3 Rearranging equations involving e^x

Make t the subject in $3x + 2 = e^{5t}$.

Solution

Take the ln of both sides: $\ln(3x+2) = \ln(e^{5t})$

$\Rightarrow \ln(3x+2) = 5t$

$\Rightarrow t = \frac{1}{5}\ln(3x+2)$

$\ln(e^{5t}) = 5t$.

4 Rearranging equations involving $\ln x$ (1)

Make a the subject in $\ln(7a + 2) = 4t$.

Solution

Raise both sides as a power of e: $e^{\ln(7a+2)} = e^{4t}$

$\Rightarrow 7a + 2 = e^{4t}$

$\Rightarrow 7a = e^{4t} - 2$

$\Rightarrow a = \frac{1}{7}(e^{4t} - 2)$

5 Rearranging equations involving $\ln x$ (2)

Make p the subject of $\ln\left(\frac{p+5}{p}\right) = 4t$.

Solution

Raise both sides as a power of e: $e^{\ln\left(\frac{p+5}{p}\right)} = e^{4t}$

$$\frac{p+5}{p} = e^{4t}$$

$$p + 5 = pe^{4t}$$

$$5 = pe^{4t} - p$$

$$5 = p(e^{4t} - 1)$$

$$p = \frac{5}{e^{4t} - 1}$$

6 Sketching graphs of exponential functions

Sketch the graph of $y = 2 + 3e^{-x}$.

Solution

When $x = 0$, $y = 2 + 3e^0 = 2 + 3 = 5$

When $x \to \infty$, $y \to 2 + 3 \times 0 = 2$

So the line $y = 2$ is a horizontal asymptote.

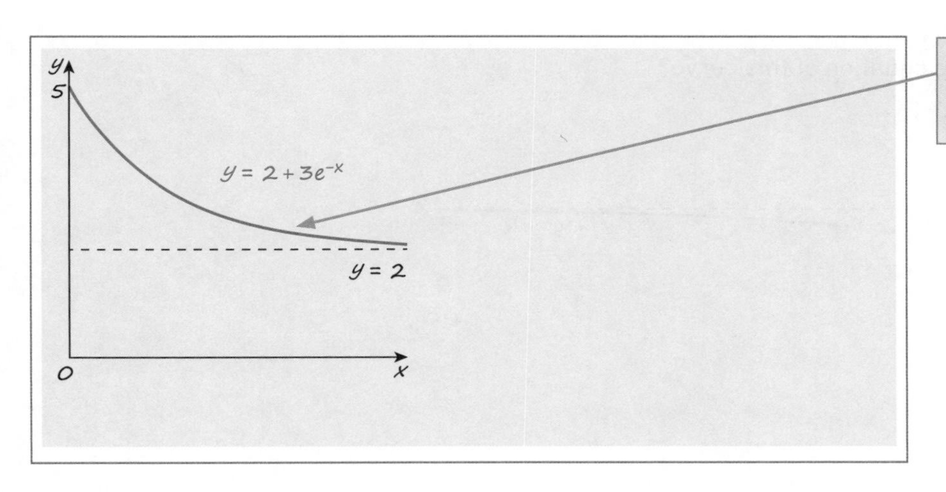

Test yourself

TESTED

1 Differentiate $y = \dfrac{6}{e^{2x}}$.

A $\dfrac{dy}{dx} = -3e^{-2x}$ B $\dfrac{dy}{dx} = -12e^{x}$ C $\dfrac{dy}{dx} = \dfrac{-1}{(12e^{2x})}$ D $\dfrac{dy}{dx} = -\dfrac{12}{e^{2x}}$ E $\dfrac{dy}{dx} = 12e^{2x}$

2 Make x the subject of the equation $\ln(3x+2) = 5t$.

A $x = \frac{1}{3}e^{(5t-2)}$ B $x = \frac{5}{3}(t - \ln 2)$ C $x = \frac{1}{3}(e^{5t} + 2)$ D $x = \frac{1}{3}e^{5t} - 2$ E $x = \frac{1}{3}(e^{5t} - 2)$

3 The graph shows the curve $y = \ln x$.

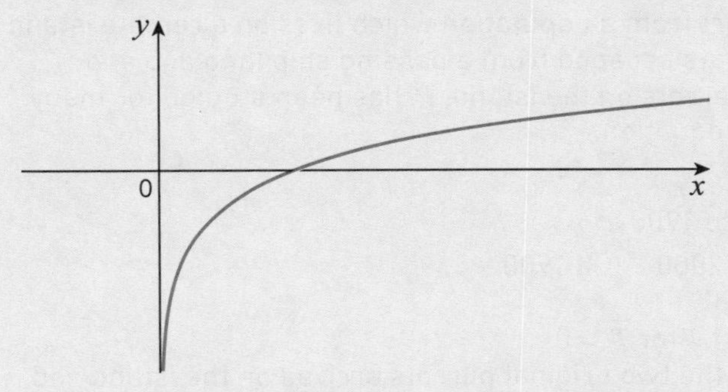

Four of the following statements are false and one is true. Which one is true?

A The graph crosses the y-axis at (0, –1).

B The graph crosses the x-axis at (e, 0).

C The graph passes through the point (e, 1).

D The graph flattens out for large values of x and approaches a horizontal asymptote.

E If you draw the graph for negative values of x, it is the same curve reflected in the y-axis.

4 You are given that $M = 100 + 300e^{-0.1t}$. Find the value of t when $M = 250$.

A 4.7 B 50.1 C 0.0693 D 6.93 E 16.5

5 Which of the following is the equation of this curve?

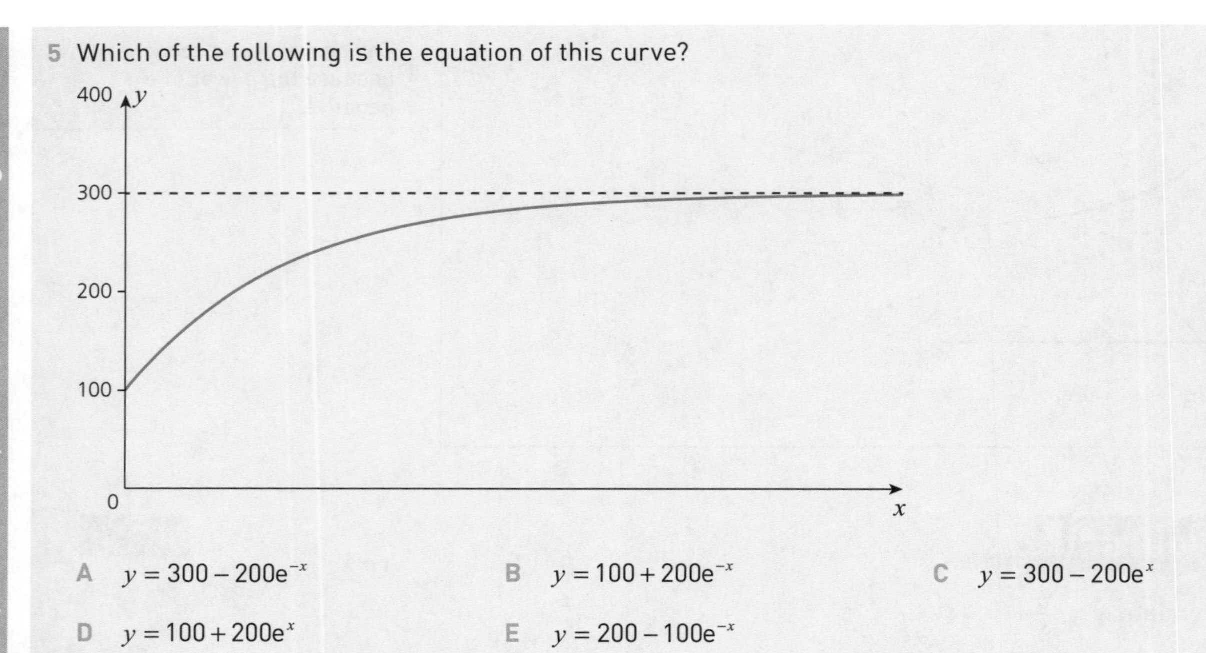

A $y = 300 - 200e^{-x}$

B $y = 100 + 200e^{-x}$

C $y = 300 - 200e^{x}$

D $y = 100 + 200e^{x}$

E $y = 200 - 100e^{-x}$

Full worked solutions online

CHECKED ANSWERS

Exam-style question

A certain type of parrot is found only in Australia apart from a population which lives on a remote island in the south Pacific. It is believed that two of the parrots escaped from a passing ship long ago and established the island's population. The number of parrots on the island, P, has been studied for many years and has been found to be well modelled by the equation:

$$P = 5000 - 3000e^{-0.008T}$$

where T is the number of years that have passed since 1900.

i Find the number of parrots on the island in **a** 2000 **b** 1900.

ii In what year will there be 4000 parrots?

iii Sketch the graph of the number of parrots against T for $T \geqslant 0$.

iv Use the equation for P to estimate the year when the two original parrots arrived on the island and give one reason why this might not be very accurate.

v Extend your graph to cover values of T less than zero.

vi Comment on the main features of the graph.

Short answers on page 157

Full worked solutions online

CHECKED ANSWERS

Modelling curves

REVISED

Key facts

Logarithms can be used to find the relationship between variables in two situations.

1 For relationships of the form $y = kx^n$, you can take logs of both sides and write:

$\log_{10} y = \log_{10} kx^n$

$\Rightarrow \log_{10} y = \log_{10} x^n + \log_{10} k$

$\Rightarrow \log_{10} y = n\log_{10} x + \log_{10} k$

and so plotting **log y against log x** gives a straight line. The gradient of the line is n and the intercept is log k.

2 For relationships of the form $y = ka^x$, you can take logs of both sides and write:

$\log_{10} y = \log_{10} ka^x$

$\Rightarrow \log_{10} y = \log_{10} a^x + \log_{10} k$

$\Rightarrow \log_{10} y = x\log_{10} a + \log_{10} k$

and so plotting **log y against x** gives a straight line. The gradient of the line is log a and the intercept is log k.

Using law: $\log ab = \log a + \log b$.

Using power law: $\log x^n = n \log x$.

Notice that this is in the form $y = mx + c$.

Worked examples

1 Plotting log y against log x

In an experiment the temperature θ°C of a cooling liquid is measured every 2 minutes. The table shows the results.

Time in minutes (t)	2	4	6	8	10
Temperature (θ)	95	65	56	38	31

i Plot the graph of $\log_{10}\theta$ against $\log_{10} t$ and draw a line of best fit.

ii Use the graph to find the relationship between θ and t.

iii Use the equation in part **ii** to predict the temperature of the liquid after 15 minutes, giving your answer to one decimal place.

iv At what time, to the nearest minute, will the temperature of the liquid be 40°C?

Solution

i Make a table to show $\log_{10} t$ and $\log_{10}\theta$ then plot the graph of $\log_{10}\theta$ against $\log_{10} t$

log t	0.30	0.60	0.78	0.90	1
log θ	1.98	1.81	1.75	1.58	1.49

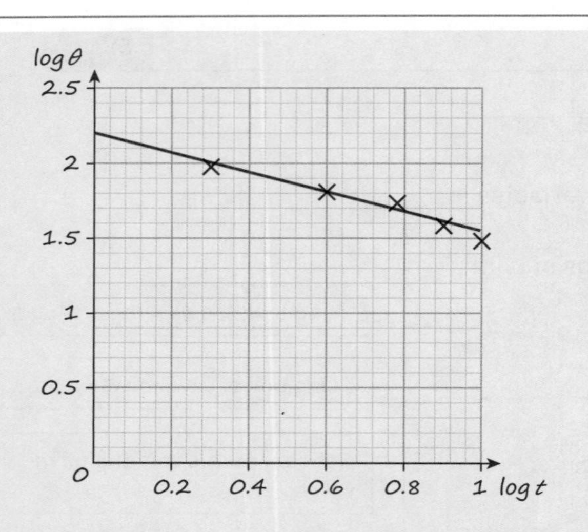

Hint: The graph is a straight line, so you can use it to find a relationship between θ and t since you know that the equation of a straight line is $y = mx + c$.

Hint: Real data will not fit a straight line exactly so you need to draw a line of best fit.

Try to draw the line of best fit as accurately as you can as this will affect your answer to the other parts of the question.

ii The graph is a straight line with an equation of the form

$$\log\theta = n\log t + \log k$$
$$\Rightarrow \log\theta = \log t^n + \log k$$
$$\Rightarrow \log\theta = \log kt^n$$
$$\Rightarrow \theta = kt^n$$

The line cuts the vertical axis at 2.2 so

$$\log_{10} k = 2.2$$
$$\Rightarrow k = 10^{2.2}$$
$$\Rightarrow k = 158.489... = 158.5 \text{ (1 d.p.)}$$

From your graph.

The line passes through the points (0, 2.2) and (1, 1.55)

The gradient of the line is:

$$\frac{2.2 - 1.55}{0 - 1} = -0.65$$
$$\Rightarrow n = -0.65$$

So the relationship between θ and t is $\theta = 158.5t^{-0.65}$

iii To find the temperature after 15 minutes, substitute $t = 15$ into the equation of the line:

$\theta = 158.489... \times 15^{-0.65} = 27.260....$

The temperature will be 27.3°C after 15 minutes.

iv To find the time at which the temperature is 40°C, substitute $\theta = 40$ into the equation of the line.

$$40 = 158.48... t^{-0.65}$$

$$\Rightarrow \quad \frac{40}{158.48}... = t^{-0.65}$$

$$\Rightarrow \log\left(\frac{40}{158.48...}\right) = \log t^{-0.65}$$

Take logarithms to base 10 of both sides.

$$\Rightarrow \quad -0.597... = -0.65 \log t$$

$$\Rightarrow \quad \frac{-0.597...}{-0.65} = \log t$$

This equals 0.919.

$$\Rightarrow \quad 10^{0.919...} = t$$

Since the inverse of $y = \log_a x$ is $y = a^x$.

$$\Rightarrow \quad 8.32 = t \text{ (to 3 s.f.)}$$

So the temperature of the liquid will be 40°C after 8 minutes (to the nearest minute).

2 Plotting log y against x

The population, P, of sparrows in a region is modelled by $P = ka^t$, where t is the time in years. The table shows the populations over a five year period.

Year (t)	1	2	3	4	5
Population	390	440	495	645	794

i Plot the graph of $\log_{10} P$ against t and use the graph to find the equation for P in terms of t.

ii Use the equation to find the population after 8 years.

iii After how long will the population be greater than 2000 according to this model?

Solution

i

t	1	2	3	4	5
$\log_{10} P$	2.59	2.64	2.69	2.81	2.90

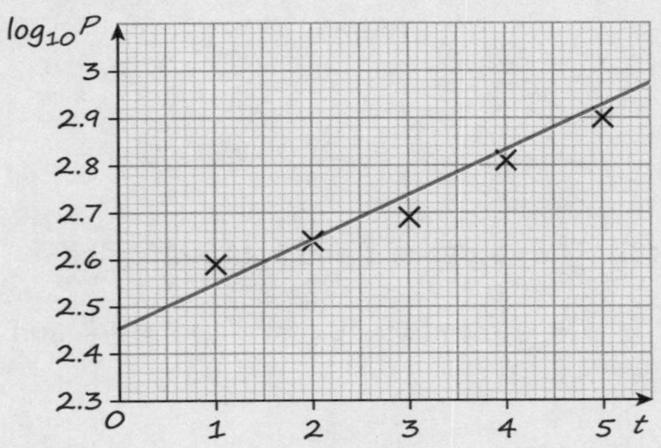

The line cuts the vertical axis at $\log_{10} k = 2.45$

So $k = 10^{2.45} \Rightarrow k = 282$ (to 3 s.f.)

The line goes through (2, 2.64) and (0, 2.45) so its gradient, log a, is given by:

$$\log_{10} a = \frac{2.64 - 2.45}{2 - 0} = \frac{0.19}{2} = 0.095$$

$\Rightarrow \quad a = 10^{0.095} = 1.24 \quad (2 \text{ d.p.})$

So the relationship between P and t is $P = 282 \times 1.24^t$

ii To find the population after 8 years, substitute $t = 8$ into the equation of the curve:

$P = 282 \times 1.24^8$

$P = 1575$ (nearest whole number)

iii To find when the population will be greater than 2000, substitute $P = 2000$ in the inequality:

$$282 \times 1.24^t > 2000$$

$$\Rightarrow \quad 1.24^t > \frac{2000}{282}$$

$$\Rightarrow \quad 1.24^t > 7.097...$$

Take logs to base 10 of both sides.

$$\Rightarrow \quad t \log 1.24 > \log 7.097$$

$$t > \frac{\log 7.097}{\log 1.24}$$

$$t > 9.11$$

So the population will be greater than 2000 after 9.11 years.

Test yourself

TESTED

1 The graph shows the result of plotting $\log_{10} y$ against x.

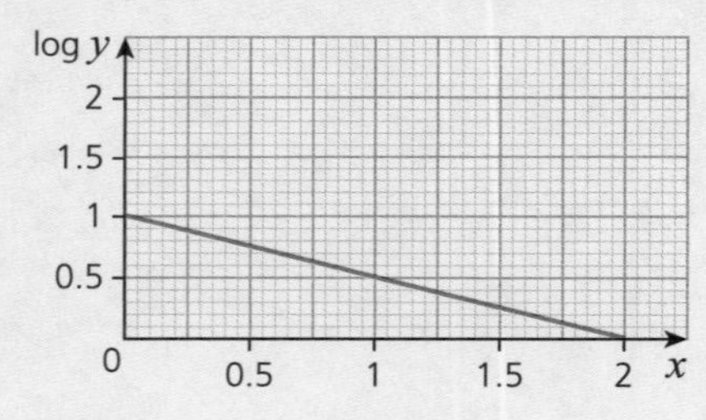

The relationship between x and y is of the form $y = k \times T^x$. The values of T and k, to 2 d.p., are

A $T = 0.32$ and $k = 10$ **B** $T = 0.32$ and $k = 1$ **C** $T = -\frac{1}{2}$ and $k = 10$ **D** $T = -\frac{1}{2}$ and $k = 1$

2 The relationship between a company's profits in thousands of Euros (P) and time in months (T) is found to be $P = 20\,T^{-0.65}$ (all numbers to 2 s.f.). Which graph represent this relationship?

A

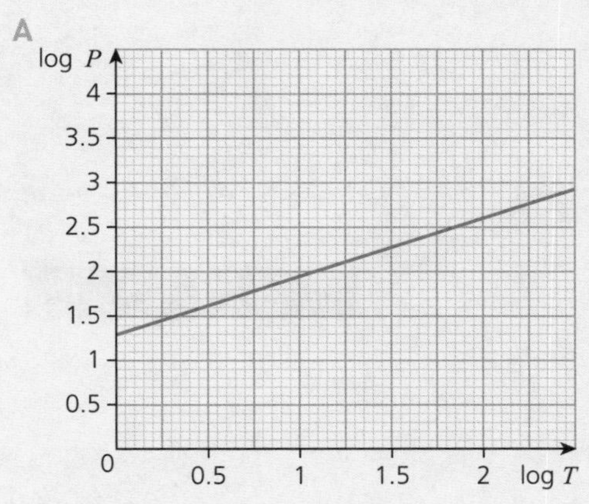

B

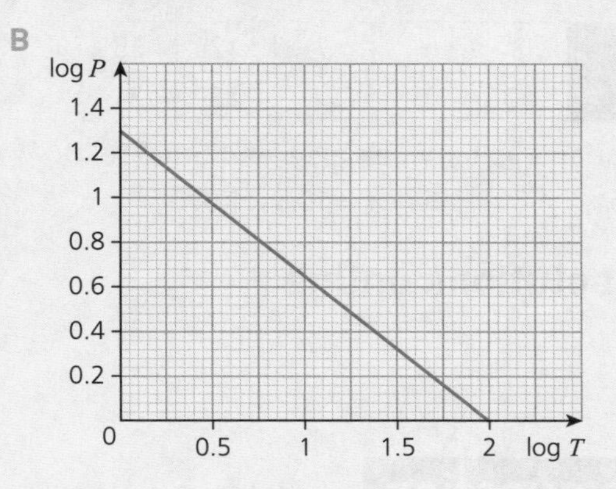

C

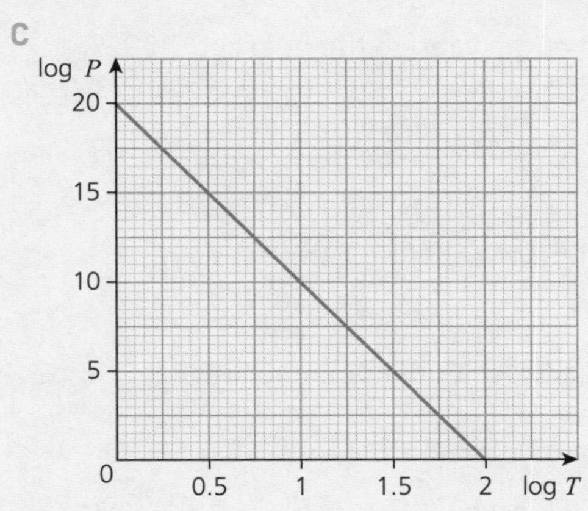

D

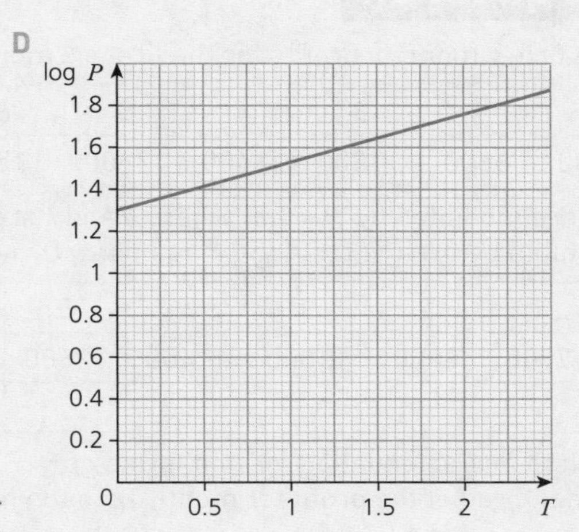

3 In an experiment, a variable, y, is measured at different times, t. The graph below shows $\log_{10} y$ against $\log_{10} t$.

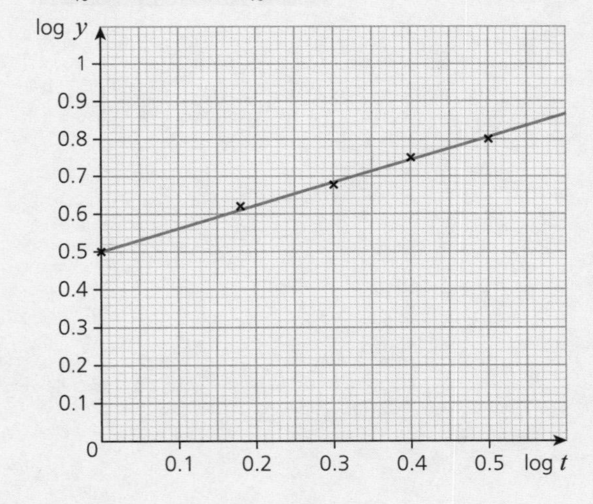

The relationship between y and t is

A $y = 3.16t^{1.67}$ B $y = 3.16t^{0.6}$ C $y = 0.5t^{0.6}$ D $y = 3.16t^{-0.22}$

4 The area of a patch of mould grows over time. Scientists measure the area, A cm², at 3 hourly intervals. The table shows the results of these measurements.

Time (hours)	3	6	9	12	15
Area (cm²)	13	19	24	28	31

What is the relationship between area and time?

A $A = 8.5 + 1.5t$ B $A = 10.5 \times 1.075^t$ C $A = \frac{1}{2} \times t^8$ D $A = 8\sqrt{t}$

Full worked solutions online

CHECKED ANSWERS

Exam-style question

The table shows a firm's monthly profits for the first six months of the year, to the nearest £100.

Month (x)	1	2	3	4	5	6
Profit (P)	7900	8800	10000	11400	12600	13500

The firm's profits are modelled by $P = ka^x$, where a and k are constants.

i Complete the table below and plot $\log_{10} P$ against x. Draw a line of best fit for the data.

Month (x)	1	2	3	4	5	6
Profit (P)	7900	8800	10000	11400	12600	13500
$\log_{10} P$						

ii Use your graph to find an equation for P in terms of x.

iii Using this model, predict the profit for month 12 to the nearest £100.

Short answers on page 158

Full worked solutions online

CHECKED ANSWERS

Review questions (Chapters 10–13)

1 A function is defined as $f(x) = 4x^3 - 9x^2 - 12x + 2$.

Find the values of x for which $f(x)$ is decreasing.

2 The equation of a curve is $y = x^3 - 3x^2 - 3$.

i Find the coordinates of the turning points and identify their nature.

ii Find the equation of the normal to the curve at the point where $x = 1$.

Give your answer in the form $ax + by + c = 0$.

3 Given $f'(x) = \frac{4}{\sqrt{x}} - \frac{1}{x^2}$, find

i $f''(x)$

ii $f(x)$

4 Find the area between the curve $y = \sqrt{x^3}$, the x-axis and the lines $x = 4$ and $x = 9$.

5 i Solve the following equations:

a $2^x = 10$ **b** $\log_{10} x + \log_{10} 4x = 3\log_{10} 4$

ii Find the value of $\log_a a^2 - \log_a \frac{1}{a}$.

6 The points A and B have position vectors $\begin{pmatrix} -3 \\ 2 \end{pmatrix}$ and $\begin{pmatrix} 5 \\ -2 \end{pmatrix}$.

i Find $|\overrightarrow{AB}|$. Give your answer as a simplified surd.

ii Find the position vector of C such that $\overrightarrow{AC} = 2\overrightarrow{BC}$.

Short answers on page 158

Full worked solutions online

CHECKED ANSWERS

SECTION 3 Review questions

Exam preparation

Before your exam

REVISED

- *Start revising early* – half an hour a day for 6 months is better than cramming in all-nighters in the week before the exam. Little and often is the key.
- *Don't procrastinate* – you won't feel more like revising tomorrow than you do today!
- Put your phone on *silent* while you revise – don't get distracted by a constant stream of messages from your friends.
- Make sure your *notes are in order* and nothing is missing.
- *Be productive* – don't waste time colouring endless revision timetables. Make sure your study time is actually spent revising!
- Don't just read about a topic. *Maths is an active subject* – you improve by answering questions and actually *doing* maths, not just reading about it.
- Time yourself on exam-style questions and check you are not spending too long on each question.
- Answer all the questions on as many past papers as you can.
- Try *teaching a friend* a topic – teaching something is the best way to learn it yourself – that's why your teachers know so much!

The exam papers

REVISED

You must take both Components 01 and 02 to be awarded the OCR AS Level in Mathematics B(MEI).

Component	Title	No. of marks		
01	**Pure Mathematics and Mechanics**	70	1 hour 30 minutes Written paper	**50%** of total AS Level
02	**Pure Mathematics and Statistics**	70	1 hour 30 minutes Written paper	**50%** of total AS Level

The content of this book covers all the Pure Mathematics content for both components.

Make sure you know these formulae for your exam

Formulae you need to know....

REVISED

Topic	Formula
Circle	*Area* $= \pi r^2$ *Circumference* $= 2\pi r$, where r is the radius
Parallelogram 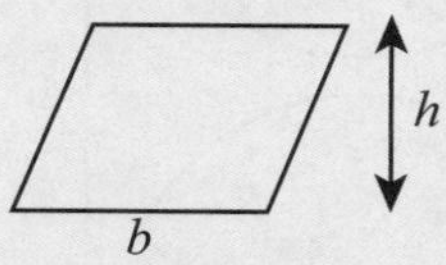	*Area = base × vertical height*
Trapezium 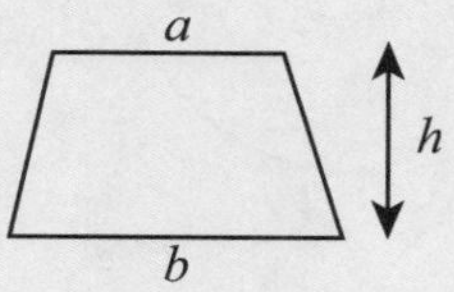	*Area* $= \frac{1}{2}h(a+b)$
Triangle 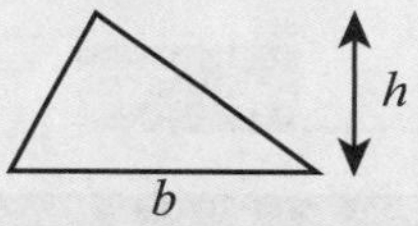	*Area* $= \frac{1}{2}$ *base × vertical height*
Prism	*Volume = area of cross section × length*
Cylinder	*Volume* $= \pi r^2 h$ *Area of curved surface* $= 2\pi rh$ *Total surface area* $= 2\pi rh + 2\pi r^2$, where r is the radius and h is the height
Pythagoras' theorem 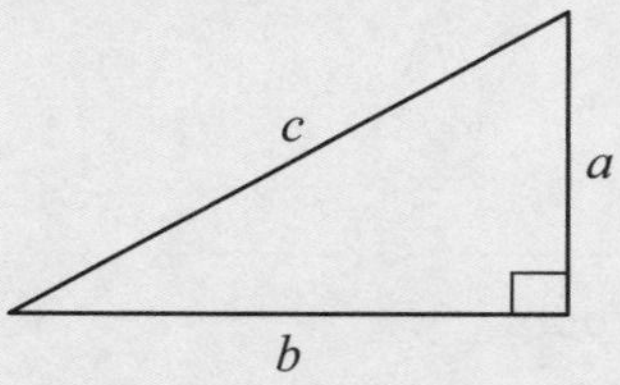	$a^2 + b^2 = c^2$
Trigonometry 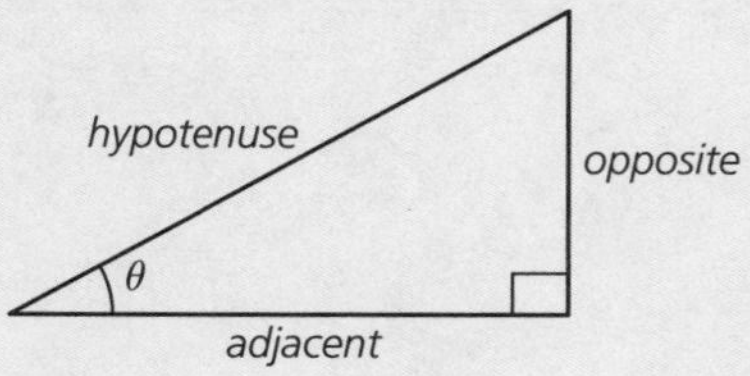	$\cos\theta = \frac{\text{adjacent}}{\text{hypotenuse}}$ $\sin\theta = \frac{\text{opposite}}{\text{hypotenuse}}$ $\tan\theta = \frac{\text{opposite}}{\text{adjacent}}$

Topic	Formula
Circle theorems	The angle in a semi-circle is a right-angle. O The perpendicular from the centre of a circle to a chord bisects the chord. O The tangent to a circle at a point is perpendicular to the radius through that point. O

From AS Pure Maths you should know ...

REVISED

Topic	Formula
Laws of indices	$a^m \times a^n = a^{m+n}$ $\frac{a^m}{a^n} = a^{m-n}$ $(a^m)^n = a^{mn}$ $a^{-n} = \frac{1}{a^n}$ $\sqrt[n]{a} = a^{\frac{1}{n}}$ $\sqrt[n]{a^m} = a^{\frac{m}{n}}$ $a^0 = 1$
Quadratic equations	The quadratic equation $ax^2 + bx + c = 0$ has roots $x = \frac{-b \pm \sqrt{b^2 - 4ac}}{2a}$.
Coordinate geometry	● For two points (x_1, y_1) and (x_2, y_2): Gradient $= \frac{y_2 - y_1}{x_2 - x_1}$ Length $= \sqrt{(x_2 - x_1)^2 + (y_2 - y_1)^2}$ Midpoint $= \left(\frac{x_1 + x_2}{2}, \frac{y_1 + y_2}{2}\right)$. ● The equation of a straight line with gradient m and y-intercept $(0, c)$ is $y = mx + c$.

Topic	Formula
	• The equation of a straight line with gradient m and passing through (x_1, y_1) is $y - y_1 = m(x - x_1)$. • Parallel lines have the same gradient. • For two perpendicular lines $m_1 m_2 = -1$. • The equation of a circle, centre (a, b) and radius r is $(x-a)^2 + (y-b)^2 = r^2$.
Trigonometry	For any triangle ABC *Area:* $\text{Area} = \frac{1}{2}ab\sin C$ *Sine rule:* $\frac{a}{\sin A} = \frac{b}{\sin B} = \frac{c}{\sin C}$ or $\frac{\sin A}{a} = \frac{\sin B}{b} = \frac{\sin C}{c}$ *Cosine Rule:* $a^2 = b^2 + c^2 - 2bc\cos A$ or $\cos A = \frac{b^2 + c^2 - a^2}{2bc}$ *Identities:* $\sin^2\theta + \cos^2\theta \equiv 1$ $\tan\theta \equiv \frac{\sin\theta}{\cos\theta}, \quad \cos\theta \neq 0$
Transformations	$y = f(x+a)$ is a translation of $y = f(x)$ by $\begin{pmatrix} -a \\ 0 \end{pmatrix}$. $y = f(x) + b$ is a translation of $y = f(x)$ by $\begin{pmatrix} 0 \\ b \end{pmatrix}$. $y = f(ax)$ is a one-way stretch of $y = f(x)$, parallel to x-axis, scale factor $\frac{1}{a}$. $y = af(x)$ is a one-way stretch of $y = f(x)$, parallel to y-axis, scale factor a. $y = f(-x)$ is a reflection of $y = f(x)$ in the y-axis. $y = -f(x)$ is a reflection of $y = f(x)$ in the x-axis.
Polynomials and binomial expansions	*The Factor theorem*: If $(x - a)$ is a factor of $f(x)$ then $f(a) = 0$ and $x = a$ is a root of the equation $f(x) = 0$. Conversely, if $f(a) = 0$ then $(x - a)$ is a factor of $f(x)$. *Pascal's triangle:* 1 1 1 1 2 1 1 3 3 1 1 4 6 4 1 *Factorials:* $n! = n \times (n-1) \times (n-2) \times \ldots \times 1$

Topic	Function	Derivative
Differentiation	$y = kx^n$	$\frac{dy}{dx} = knx^{n-1}$
	$y = e^{kx}$	$\frac{dy}{dx} = ke^{kx}$
	$y = f(x) + g(x)$	$\frac{dy}{dx} = f'(x) + g'(x)$

Topic	Formula	
Integration	***Function:*** $\int kx^n \mathrm{d}x$	***Integral:*** $\frac{kx^{n+1}}{n+1} + c$
Vectors	$\overrightarrow{AB} = \overrightarrow{OB} - \overrightarrow{OA}$ If $\mathbf{a} = x\mathbf{i} + y\mathbf{j}$ then $\|\mathbf{a}\| = \sqrt{x^2 + y^2}$	
Exponentials and logarithms	$y = \log_a x \Leftrightarrow a^y = x$ for $a > 0$ and $x > 0$ $\log xy = \log x + \log y$ $\quad \log \sqrt[n]{x} = \frac{1}{n}\log x$ $\quad \log_a a = 1$ $\log \frac{x}{y} = \log x - \log y$ $\quad \log \frac{1}{x} = -\log x$ $\quad \mathrm{e} = 2.718...$ $\log x^n = n \log x$ $\quad \log 1 = 0$ $\quad \log_e x = \ln x$	

Formulae that will be given

REVISED

Make sure you are familiar with the formula book you will use in the exam. The formulae sheet is subject to change, so always check the latest version on the exam board's website.

Here are the formulae you are given for the **AS Pure Mathematics** part of the exam.

Topic	Formula
Binomial series	$(a+b)^n = a^n + {}^nC_1 a^{n-1}b + {}^nC_2 a^{n-2}b^2 + ... + {}^nC_r a^{n-r}b^r + ... + b^n ...(n \in \mathbb{N})$, where ${}^nC_r = \binom{n}{r} = \frac{n!}{r!(n-r)!}$ ${}_nC_r$ is commonly used as an alternative to ${}^nC_r$. $(1+x)^n = 1 + nx + \frac{n(n-1)}{2!}x^2 + ... + \frac{n(n-1)...(n-r+1)}{r!}x^r + ...(\|x\| < 1, n \in \mathbb{R})$ This is on your AS formula sheet but you only need this for A level Maths.
Differentiation from first principles	$\mathrm{f}'(x) = \lim_{h \to 0} \frac{\mathrm{f}(x+h) - \mathrm{f}(x)}{h}$

If you are studying A Level Mathematics you will be given a longer formula sheet. See the A level Pure Mathematics Revision Guide.

During your exam

Watch out for these key words

REVISED

- **Exact** … leave your answer as a simplified surd, fraction or power.

  Examples: $\ln 5$ ✓ 1.61 ✗ e^2 ✓ 7.39 ✗
  $2\sqrt{3}$ ✓ 3.26 ✗ $1\frac{5}{6}$ ✓ 1.83 ✗

- **Give/State/Write down** … no working is expected – unless it helps you.
  The marks are for the answer rather than the method.

  Example: The equation of a circle is $(x + 2)^2 + (y - 3)^2 = 13$
  Write down the radius of the circle and the coordinates of its centre.

Make sure you give both answers!

- **Prove/Show that** … the answer has been given to you. You must show full working otherwise you will lose marks. Often you will need the answer to this part to answer the next part of the question. Most of the marks will be for the method.

  Example: **i Prove** that $\sin x - \cos^2 x \equiv \sin^2 x + \sin x - 1$

- **Hence** ... you **must** follow on from the given statement or previous part. Alternative methods may not earn marks.

  Example: ii Hence solve $\sin x - \cos^2 x = -2$ for $0° \leqslant x \leqslant 180°$

Remember that if you couldn't answer part i you can still go on and answer part ii.

- **Hence or otherwise** ... there may be several ways you can answer this question but it is likely that following on from the previous result will be the most efficient and straightforward method.

  Example: Factorise $p(x) = 6x^2 + x - 2$
  Hence, or otherwise, solve $p(x) = 0$

Watch out for these common mistakes

REVISED

- ✗ Miscopying your own work or misreading/miscopying the question.
- ✗ Not giving your answer as coordinates when it should be e.g. when finding where two curves meet.
- ✗ Giving your answer as coordinates inappropriately e.g. writing a vector as coordinates.
- ✗ Not finding y coordinates when asked to find coordinates.
- ✗ Not finding where the curve cuts the x **and** y axes when sketching a curve.
- ✗ Using a ruler to draw curves.
- ✗ Not using a ruler to draw straight lines.
- ✗ Spending too long drawing graphs when a sketch will do.
- ✗ Not stating the equations of asymptotes when sketching an exponential or reciprocal curve.
- ✗ Not simplifying your answer sufficiently.
- ✗ Rounding answers that should be **exact**.
- ✗ Rounding errors – don't round until you reach your final answer.

- ✗ Giving answers to the wrong degree of accuracy – use 3 s.f. unless the questions says otherwise.
- ✗ Not showing any or not showing enough working – especially in 'show that' or 'proof' questions.

Once you have answered a question **re-read the question** making sure you've answered **all** of it. It is easy to miss the last little bit of a question.

Check your answer is ...

REVISED

- ✓ To the correct accuracy
- ✓ in the right form
- ✓ complete ... have you answered the whole question?

If you do get stuck ...

REVISED

Keep calm and don't panic.

- ✓ **Reread the question** ... have you skipped over a key piece of information that would help? Highlight any numbers or key words.
- ✓ **Draw** ... a diagram. This often helps. Especially in questions on graphs, coordinate geometry and vectors, a sketch can help you see the way forward.
- ✓ **Look** ... for how you can re-enter the question. Not being able to answer part **i** doesn't mean you won't be able to do part **ii**. Remember the last part of a question is not necessarily harder.
- ✓ **Move on** ... move onto the next question or part question. Don't waste time being stuck for ages on one question, especially if it is only worth one or two marks.
- ✓ **Return later** ... in the exam to the question you are stuck on – you'll be surprised how often inspiration will strike!
- ✓ **Think positive!** You are well prepared, believe in yourself!

Good luck!

Answers

Here you will find the answers to the 'Target your revision' exercises, 'Exam-style questions' and 'Review questions' in the book. Full worked solutions for all of these (including 'show' questions that do not have short answers in the book) are available online at www.hoddereducation.co.uk/myrevisionnotesdownloads. Note that answers for the 'Test yourself' multiple choice question are available online only.

SECTION 1

Target your revision (page 1)

1 $P \Leftarrow Q$

4 When $n = 4$ then $n^2 - 8n + 15 = -1$

5 $6\sqrt{3}$

6 $7 - 4\sqrt{3}$

7 $\dfrac{3c^2}{b}$

8 $\dfrac{7}{2}\sqrt{2}$

9 $x = -\dfrac{1}{3}$ or $x = \dfrac{3}{2}$

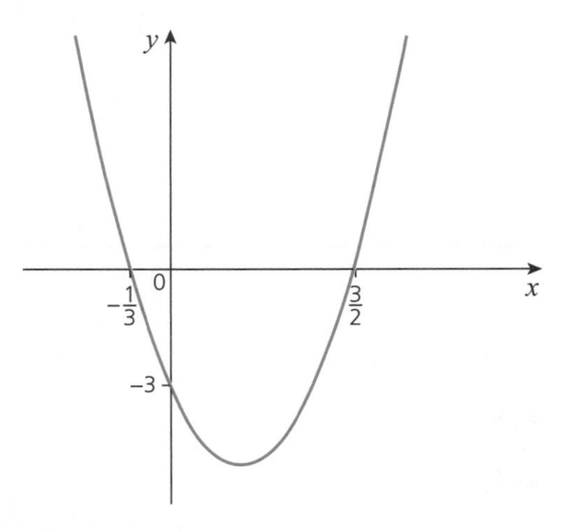

10 $y = (x - 2)^2 - 7$

11 $k = \pm 12$

12 $(-1, -1)$ and $\left(\dfrac{3}{2}, 4\right)$

13 $-2 \leqslant x < \dfrac{3}{2}$

14 $x < -3$ or $x > 5$ which can be written as

$\{x : x < -3\} \cup \{x : x > 5\}$

15

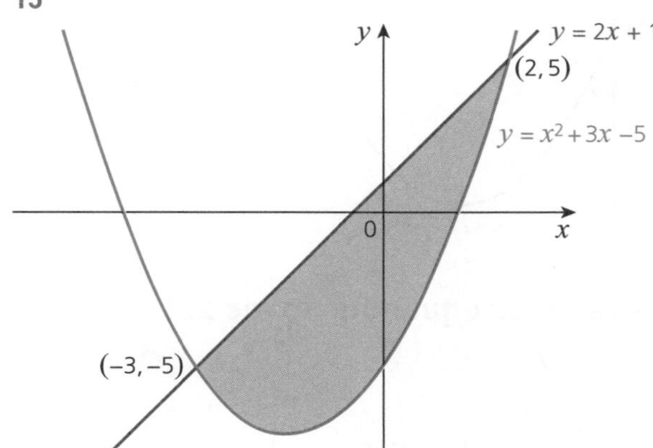

Problem solving (page 5)

i $n + 1, n + 2, n + 3, n + 4$

ii $5(n + 2)$ is divisible by 5

iii 95

Surds and indices (page 9)

i $a = 2$ and $b = \dfrac{1}{4}$

ii $a = 9$ and $b = 10$

Quadratic equations (page 13)

i

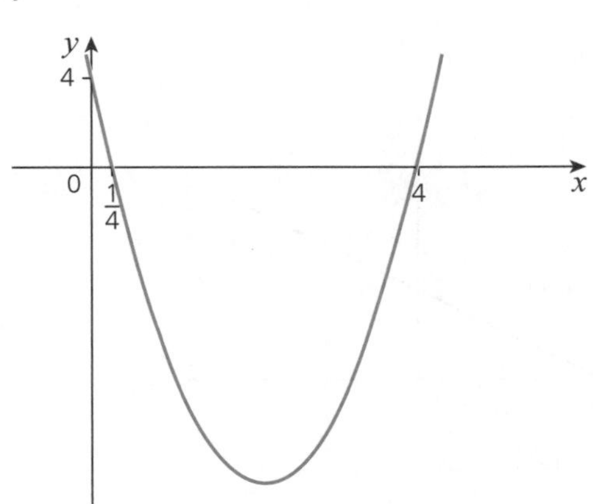

ii a $x = \pm 2$ or $x = \pm\frac{1}{2}$

b $x = \frac{1}{16}$ or $x = 16$

Completing the square and the quadratic formula (page 18)

i $f(x) = 3(x-2)^2 - 18$

ii a P(0, −6) and Q(2, −18)

b

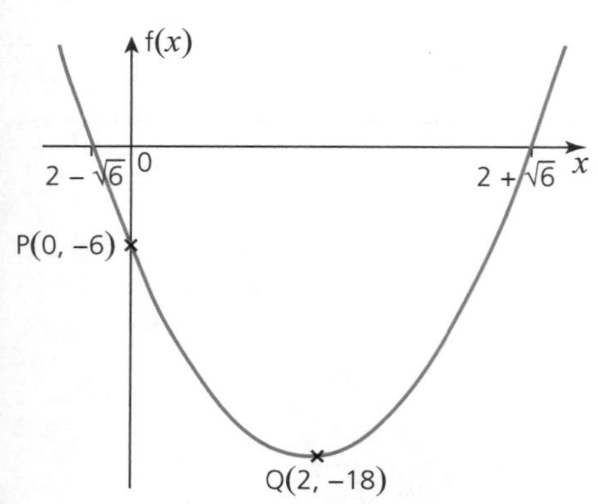

Simultaneous equations (page 22)

i $2x + y = 34$,

$xy = 144$

iii Length 18 m and width 8 m, or length 16 m and width 9 m.

Inequalities (page 27)

i $x \leqslant -2$

ii $x < -2$ or $x > 3$

Review questions (page 28)

1 i $P \Leftarrow S$

ii $S \Rightarrow Q$

iii $R \Leftrightarrow S$

3 i $(2x-1)(x+3)$

ii

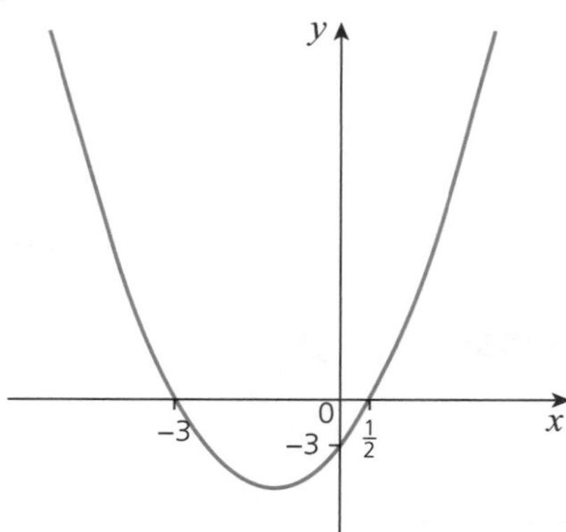

4 i $(11 - 6\sqrt{2})\text{cm}^2$

ii $c = 4$

5 i $x = 2$ and $y = 3$

ii $x = \pm 1, x = \pm 2$

6 i $f(x) = (x-5)^2 - 21$

ii $-8 < k < 8$

SECTION 2

Target your revision (page 29–30)

1 i (−2, 7)

ii $-\frac{1}{2}$

iii $4\sqrt{5} = 8.94$ to 3 s.f.

2 i $y = \frac{1}{2}x + 2$

ii

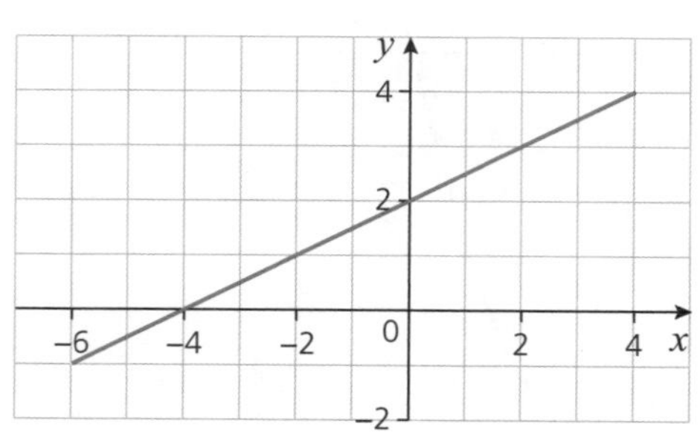

3 −0.1

4 $\left(\frac{2}{3}, -\frac{1}{3}\right)$

5 i $(x-2)^2 + (y+3)^2 = 16$

ii $(x-1)^2 + (y-3)^2 = 13$

6 $\left(-\frac{2}{3}, -\frac{19}{9}\right)$ and (2, 5)

7 $3y + 4x = 23$

8 i $-\frac{\sqrt{3}}{2}$

ii $-\frac{1}{2}$

iii 1

9 i 60°, 120°

ii 20°, 40°, 140°, 160°

iii 15°, 75°

11 i 12 cm²

ii 4.11 cm

12 i $5x^3 + 2x^2 - 5x + 1$

ii $5x^3 - 2x^2 + x + 5$

iii $10x^5 - 15x^4 - 14x^3 + 12x^2 - 5x - 6$

iv $5x^2 - 5x + 3$

13 iii $f(x) = (2x - 1)(x + 2)(x - 1)$;

$x = -2,\ x = \frac{1}{2},\ x = 1$

iv

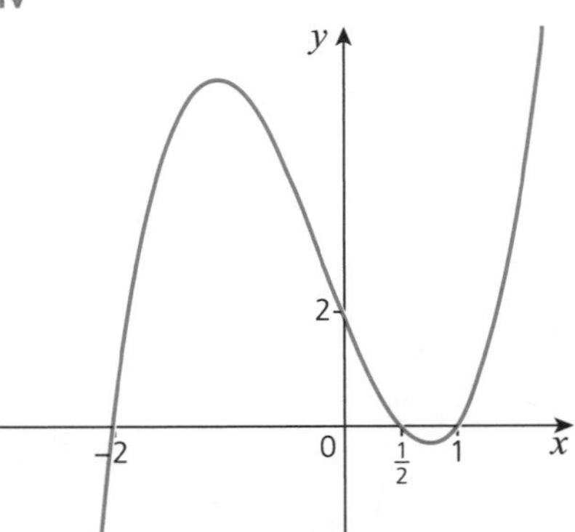

14 $y = \frac{16}{x^3}$

15 i

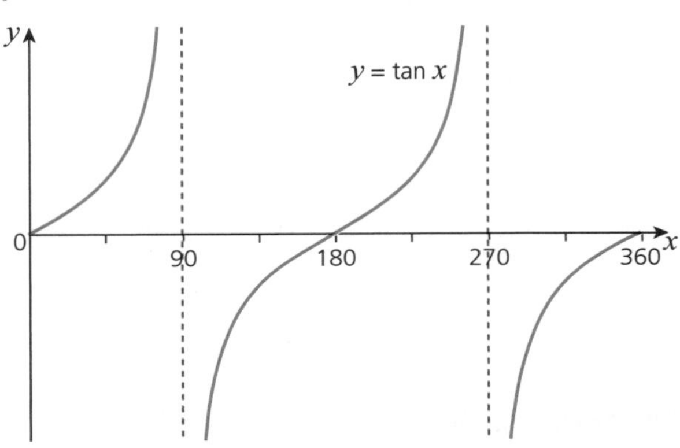

ii

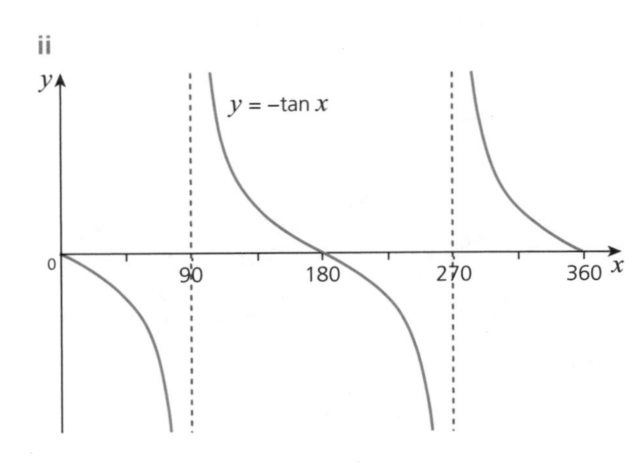

iii

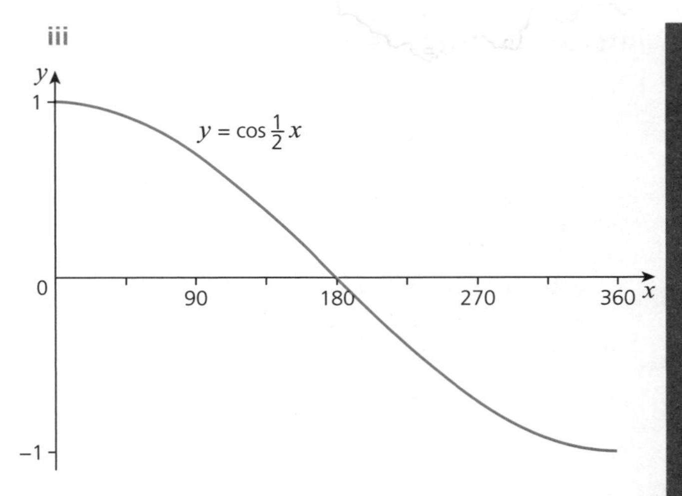

iv

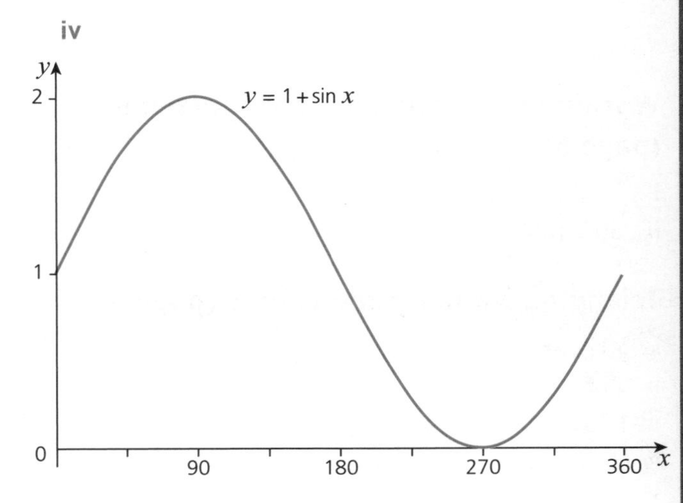

16 i $81x^4 - 216x^3 + 216x^2 - 96x + 16$

ii −2 099 520

Working with coordinates (page 35)

i (4, 3)

ii AB = $4\sqrt{5}$; CM = $2\sqrt{5}$

iii 20 square units

The equation of a straight line (page 39)

i (1, 4)

ii $1\frac{2}{3}$ square units

The circle (page 44)

ii $(5 + \sqrt{21},\ 0), (5 - \sqrt{21},\ 0)$

iii $y = -\frac{3}{4}x + 8$

Intersections (page 49)

i (1, 0), (−3, 0) and (0, −3)

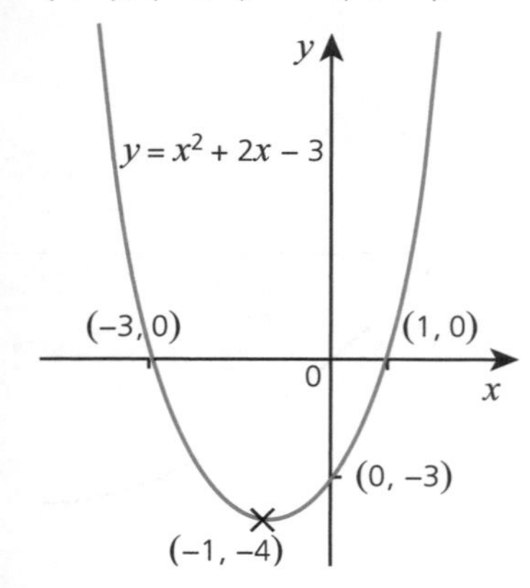

ii (2, 5)

Working with trigonometric functions (page 55)

i $-2\sin^2 x - 3\sin x + 2$

ii 30°, 150°

Triangles without right angles (page 59)

i 5.12 cm

ii 45.5°

iii 1.98 cm

iv 15.7 cm²

Polynomial expressions and polynomial curves (page 65)

ii

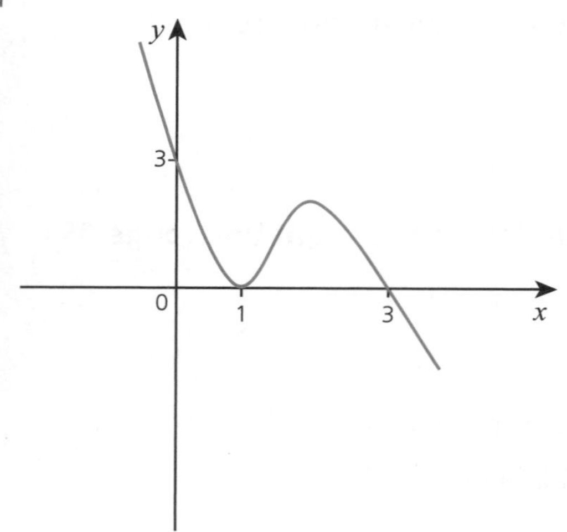

Dividing polynomials (page 67)

$x^3 - x^2 - 2x + 5$; $d = 5$

The factor theorem (page 71)

$a = -9$, $b = 9$

Curve sketching and transformations (page 78)

i

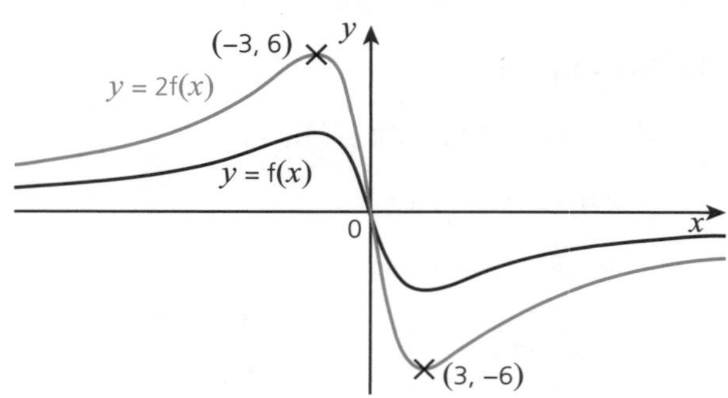

ii

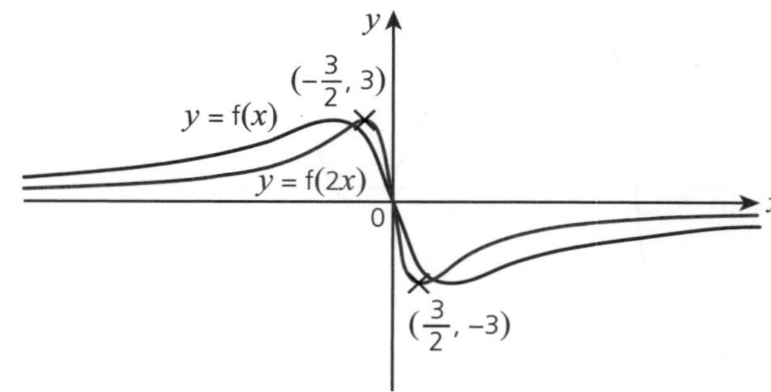

iii

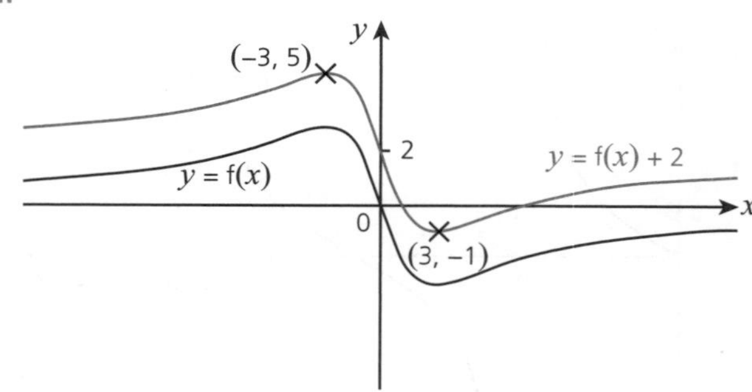

Transformations and graphs of trigonometric functions (page 81)

i

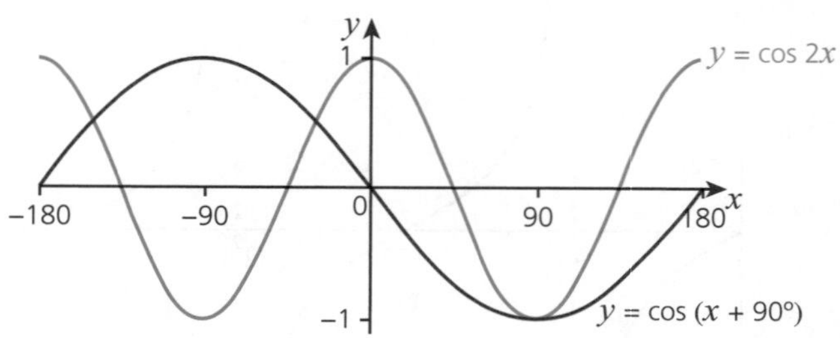

ii 3 roots

iii $x = -150°$, $x = -30°$ and $x = 90°$

iv 2

Binomial expansions and selections (page 85)

i a $1 - 20x + 180x^2 - 960x^3 + 3360x^4 + ...$

b 0.817

ii −1440

Review questions (page 86)

1 i AB = $2\sqrt{5}$, BC = $\sqrt{65}$

ii 15 square units

2 i $2y+5x=24$

ii $(x+3)^2+(y-5)^2=116$

3 ii 60°, 300°

4 i Asymptotes $x=0$ and $y=0$

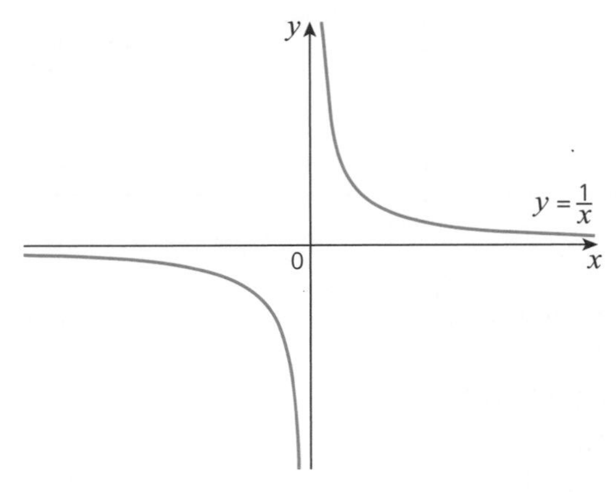

ii Asymptotes $x=0$ and $y=2$

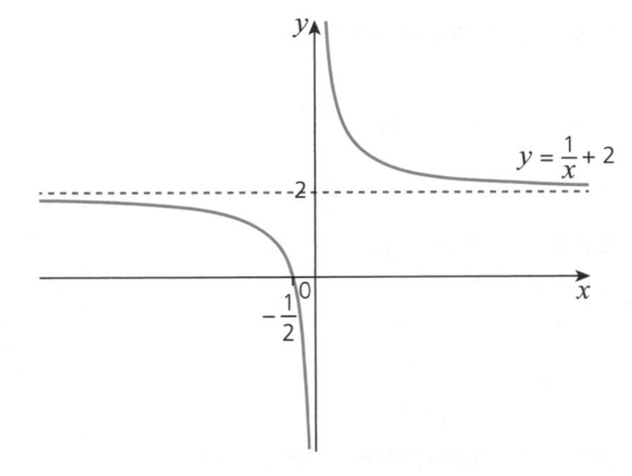

iii Asymptotes $x=-2$ and $y=0$

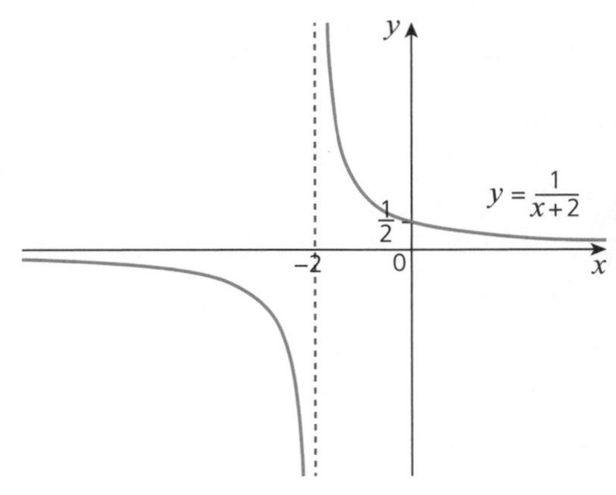

5 $n=4$, $b=-54$, $c=\frac{27}{2}$

6 i $f(x)=(x+3)(2x+1)(x-4)$

ii $x=-1$, $x=\frac{3}{2}$, $x=6$

SECTION 3

Target your revision (page 87–8)

1 i $\frac{dy}{dx}=6x-2$

ii $\frac{dy}{dx}=\frac{1}{2\sqrt{x}}-\frac{4}{x^3}$

2 i 5.5

ii 4.5

3 i $y=12x-27$

ii $y=12-2x$

4 maximum: (2, 0); minimum: $\left(\frac{2}{3},-\frac{32}{27}\right)$

5 $x<-3$, $x>2$

6

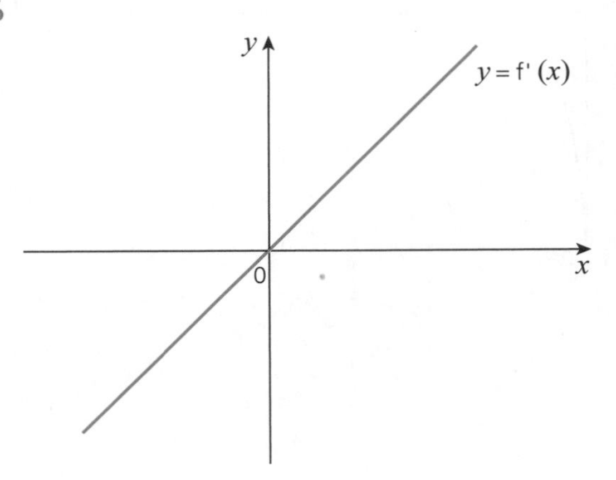

Answers

7 $\frac{3}{64}$

8 i $x^2 + 2hx + h^2$

ii $10hx + 5h^2$

iii $\frac{dy}{dx} = 10x$

9 i $x^4 - x^2 + 3x + c$

ii $\frac{2\sqrt{x^3}}{3} + \frac{3}{x} + c$

10 i $\frac{35}{3}$

ii 37

11 i 15.75 square units

ii $5\frac{1}{3}$ square units

iii $21\frac{1}{12}$ square units

iv $10\frac{5}{12}$ square units

12 $\begin{pmatrix} -4 \\ 6 \end{pmatrix}$

13 i $\begin{pmatrix} 4 \\ 2 \end{pmatrix}$

ii $2\sqrt{10}$

iii 10 square units

14 i

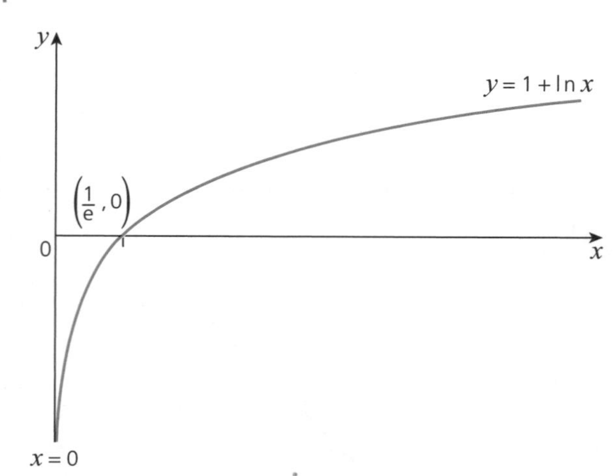

ii

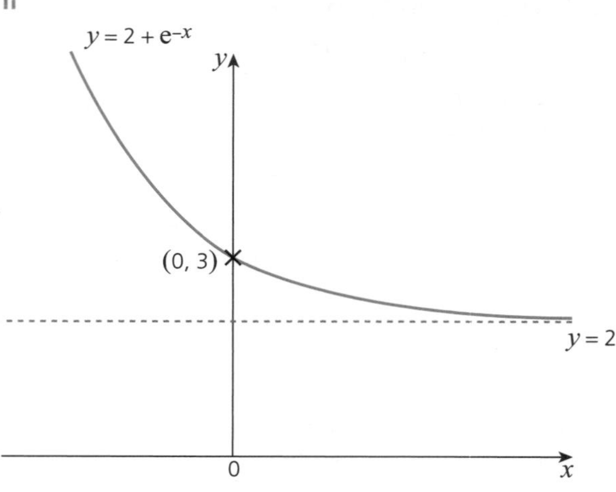

15 $\log 2\sqrt{x}$

16 i 8.68 to 3 s.f.

ii 20

17 ii $A = 100$, $k = 3$

iii 7.94×10^8

iv −0.100 to 3 s.f.

Finding gradients (page 92)

i 5.2

ii Any value between 2 and 2.1

iii 5

Extending the rule (page 94)

i $\frac{dy}{dx} = 1 + \frac{12}{x^5} - \frac{10}{x^6}$

ii $\frac{dy}{dx} = 6x - 5\sqrt{x^3} - \frac{1}{3\sqrt[3]{x^4}}$

Tangents and normals (page 98)

ii Q(−1, 2)

iii $3y = -x + 5$

Increasing and decreasing functions, and turning points (page 102)

i $\frac{dy}{dx} = 3x^2 - 6x - 9$

ii $x < -1$ or $x > 3$

iii maximum: (−1, 7); minimum: (3, −25)

Higher derivatives and the graph of $\frac{dy}{dx}$ (page 107)

i

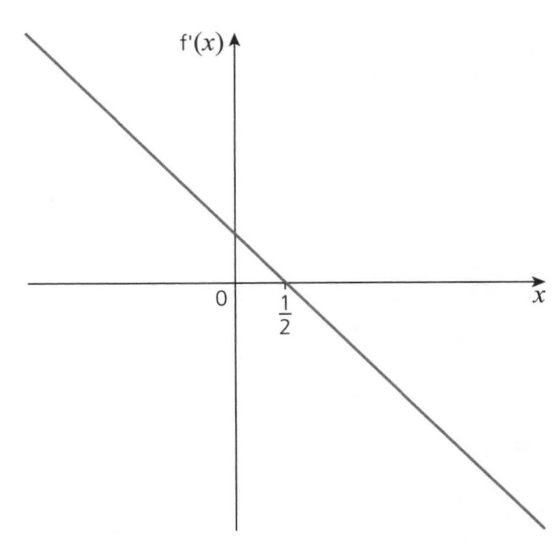

ii $x = \frac{1}{6}$; minimum

Applications and differentiation from first principles (page 111)

2670 cm^2

Integration as the reverse of differentiation (page 115)

i $2x^3 + \frac{5x^2}{2} - 3x + c$

ii $y = x^3 - 2x + 5$

Finding areas (page 119)

i A(−4, 0); B(4, 0)

ii 32 square units

iii £240

Extending the rule (page 123)

i $2(\sqrt{x})^3 - \frac{1}{x} + c$

ii $y = \frac{5}{3} - \frac{2}{x}$

Working with vectors (page 127)

i $\begin{pmatrix} 2 \\ -8 \end{pmatrix}$

ii $2\sqrt{17}$, 284°

iii a $\begin{pmatrix} 1 \\ -4 \end{pmatrix}$

b $\begin{pmatrix} 3 \\ -1 \end{pmatrix}$

Exponential functions and logarithms (page 132)

i a $3\log_a p$

b A $p = 1000$

B $p = 8$

ii $z = 3$

Natural logarithms and exponentials (page 136)

i a 3652 parrots

b 2000 parrots

ii 2037

iii

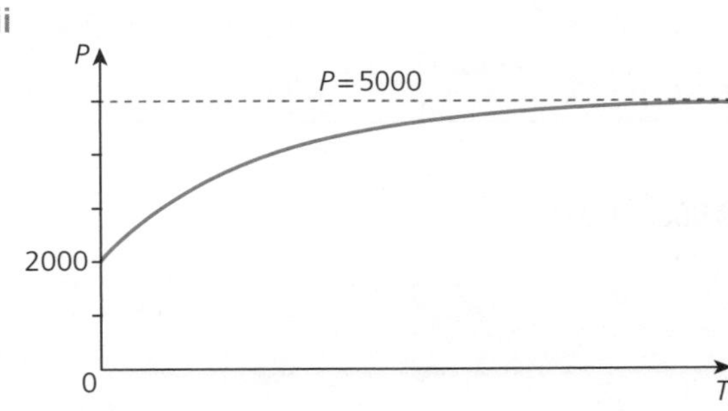

iv 1836

The model was derived from recent studies when the population is over 3000 parrots; it might not have applied when there were many fewer parrots.

There are no data for that time.

Predators and disease might have been more, or less, significant then.

v

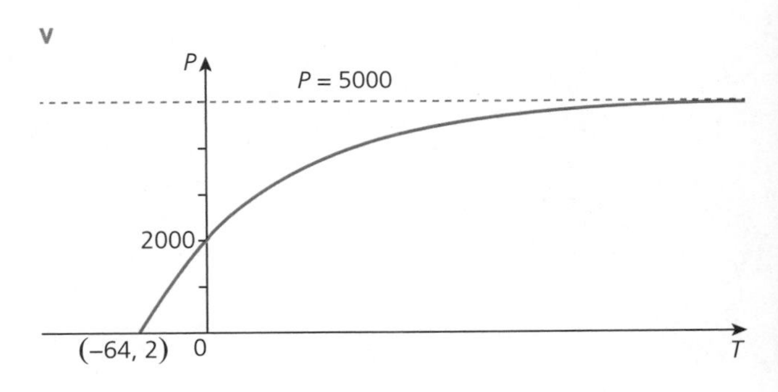

vi For large negative values of T, the graph does not exist. There were no parrots on the island.

The graph begins at (−64, 2) when the first two parrots arrived

When $T=0$, corresponding to the year 1900, the value of P is 2000.

For large positive values of T, the population P approaches the asymptote of 5000.

Modelling curves (page 142)

i

Month (x)	1	2	3	4	5	6
Profit (P)	7900	8800	10000	11400	12600	13500
$\log_{10} P$	3.90	3.94	4	4.06	4.10	4.13

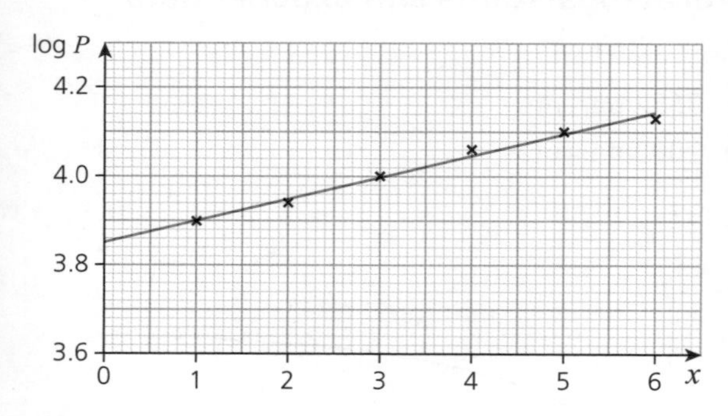

ii $P = 7080 \times 1.12^x$

iii £28200

Review questions (page 143)

1 $-\frac{1}{2} < x < 2$

2 i maximum (0, −3); minimum (2, −7)

ii $-x + 3y + 16 = 0$

3 i $\frac{2}{x^3} - \frac{2}{\sqrt{x^3}}$

ii $8\sqrt{x} + \frac{1}{x} + c$

4 84.4 square units.

5 i a $x = 3.32$ to 3 s.f.

b $x = 4$

ii 3

6 i $4\sqrt{5}$

ii $\begin{pmatrix} 13 \\ -6 \end{pmatrix}$